AF880306

Signals and Communication Technology

Series Editors

Emre Celebi, Department of Computer Science, University of Central Arkansas, Conway, AR, USA

Jingdong Chen, Northwestern Polytechnical University, Xi'an, China

E. S. Gopi, Department of Electronics and Communication Engineering, National Institute of Technology, Tiruchirappalli, Tamil Nadu, India

Amy Neustein, Linguistic Technology Systems, Fort Lee, NJ, USA

Antonio Liotta, University of Bolzano, Bolzano, Italy

Mario Di Mauro, University of Salerno, Salerno, Italy

This series is devoted to fundamentals and applications of modern methods of signal processing and cutting-edge communication technologies. The main topics are information and signal theory, acoustical signal processing, image processing and multimedia systems, mobile and wireless communications, and computer and communication networks. Volumes in the series address researchers in academia and industrial R&D departments. The series is application-oriented. The level of presentation of each individual volume, however, depends on the subject and can range from practical to scientific.

Indexing: All books in "Signals and Communication Technology" are indexed by Scopus and zbMATH.

For general information about this book series, comments or suggestions, please contact Mary James at mary.james@springer.com or Ramesh Nath Premnath at ramesh.premnath@springer.com.

Seokwon Yeom
Editor

4th International Conference on Electronics and Signal Processing

ICESP 2023

Springer

Editor
Seokwon Yeom
Daegu University Gyeongsan
Gyeongsan-si, Korea (Republic of)

ISSN 1860-4862 ISSN 1860-4870 (electronic)
Signals and Communication Technology
ISBN 978-3-031-36669-7 ISBN 978-3-031-36670-3 (eBook)
https://doi.org/10.1007/978-3-031-36670-3

© The Editor(s) (if applicable) and The Author(s), under exclusive license to Springer Nature Switzerland AG 2024
This work is subject to copyright. All rights are solely and exclusively licensed by the Publisher, whether the whole or part of the material is concerned, specifically the rights of translation, reprinting, reuse of illustrations, recitation, broadcasting, reproduction on microfilms or in any other physical way, and transmission or information storage and retrieval, electronic adaptation, computer software, or by similar or dissimilar methodology now known or hereafter developed.
The use of general descriptive names, registered names, trademarks, service marks, etc. in this publication does not imply, even in the absence of a specific statement, that such names are exempt from the relevant protective laws and regulations and therefore free for general use.
The publisher, the authors, and the editors are safe to assume that the advice and information in this book are believed to be true and accurate at the date of publication. Neither the publisher nor the authors or the editors give a warranty, expressed or implied, with respect to the material contained herein or for any errors or omissions that may have been made. The publisher remains neutral with regard to jurisdictional claims in published maps and institutional affiliations.

This Springer imprint is published by the registered company Springer Nature Switzerland AG
The registered company address is: Gewerbestrasse 11, 6330 Cham, Switzerland

Paper in this product is recyclable.

Conference Committees

Conference General Chair
James Tin-Yau Kwok, The Hong Kong University of Science and Technology, Hong Kong, China

Local Organizing Chair
Chi Man Pun, University of Macau, Macau, China

Conference General Co-Chair
LEUNG Andrew C S, City University of Hong Kong, Hong Kong, China

Conference Program Chairs
Hong Lin, University of Houston-Downtown, USA
Kenneth K.Y. Wong, University of Hong Kong, Hong Kong, China

Conference Program Co-Chair
Xiao Zhou, Wuhan University of Technology, China

Publicity Chair
Seokwon Yeom, Daegu University Gyeongsan, South Korea

Technical Program Committees
Lu Tang, Southeast University, China
Chawalit Benjangkaprasert, King Mongkut's Institute of Technology Ladkrabang, Thailand
Xiaochun Cheng, Middlesex University, UK
Lorna Uden, Staffordshire University, UK
Luis Gómez, Universidad de Las Palmas de Gran Canaria, Spain
Pavlo Maruschak, Ternopil Ivan Puluj National Technical University, Ukraine
Juan Suardíaz Muro, Universidad Politécnica de Cartagena, Spain
Azizi Abdullah, University Kebangsaan Malaysia, Malaysia

Sompong Liangrocapart, Mahanakorn University of Technology, Thailand
Vinh Truong Hoang, Ho Chi Minh City Open University, Vietnam
Suphongsa Khetkeeree, Mahanakorn University of Technology, Thailand
Thaweesak Yingthawornsuk, King Mongkut's University of Technology Thonburi, Thailand
Sheli Sinha Chaudhuri, Jadavpur University, India
Aruna Bhat, Delhi Technological University, India
Ying-Hao Yu, National Chung Cheng University, Taiwan, China
Ghufran Ahmed, National University of Computer & Emerging Sciences (NUCES), Pakistan

Preface

It is our great pleasure to welcome you to 2023 4th International Conference on Electronics and Signal Processing (ICESP 2023) which supposed to be held in Macau, China, during January 13–15, 2023, but changed to be held online due to COVID-19 pandemic. ICESP 2023 is dedicated to issues related to *Electronics and Signal Processing*. The conference was co-sponsored by the University of Macau, supported by Wuhan University of Technology, Daegu University, Mahanakorn University of Technology, etc. Again, the online conference was of great success, totally beyond anybody's expectation, having a diverse group of expertise and backgrounds. Not to mention that participants are geographically diverse. We are glad to confirm that such diverse expertise and backgrounds were connected, forming a coherent conference environment and creating synergistic learning and sharing.

ICESP 2023 is the premier forum for the presentation of new advances and research results in the fields of theoretical, experimental, and practical electronics and signal processing. The conference indeed brought together leading researchers, engineers, and architects in the domain of interest from around the world. ICESP 2023 was composed of 2 oral parallel sessions (15 min. for each talk, including 2–3 min. for Q&A), 4 keynote speeches (45 min. for each keynote speech, including 5 min. for Q&A) delivered, respectively, by Prof. James Tin-Yau Kwok, IEEE Fellow, The Hong Kong University of Science and Technology, Hong Kong, China; Prof. Chi Man Pun, University of Macau, Macau, China; Prof. Hong Lin, University of Houston-Downtown, Houston, Texas, USA; and Prof. Leung Andrew CS, City University of Hong Kong, Hong Kong, China. And finally, we would like to thank all of our conference committee members and participants for always being supportive to the conference and joining online during their busy schedule to share valuable knowledge with us (and despite the time zone difference). As expected, the conference with all the keynote speeches and presentations was proven to be intellectually stimulating to all.

We just regret that we would have been in Macau, China, but circumstances have made our conference virtual. Nevertheless, we still have internationally spread presenters that justifies saying International. If you find this volume of proceedings interesting while you're reading it, maybe you can submit a paper to ICESP next year. If you have presented a paper this year, you can also come again and help the conference grow. Here is hoping we can all meet in person next year in Macau, China, during January 12–14, 2024.

Looking forward to meeting you always at our ICESP conference.

Gyeongsan-si, Korea (Republic of) Seokwon Yeom

Contents

Part I
Data Model and Image Analysis

Two-Dimensional Imaging of Sea-Surface Targets in the Terahertz Band

Xiaofan Li, Bin Deng, Qiang Fu, and Hongqiang Wang

1 Introduction

Commonly, corner reflectors include dihedral corner reflectors, trihedral corner reflectors, octahedral corner reflectors, and icosahedral corner reflectors. The intersecting surfaces of the reflectors are perpendicular to each other, which can reflect the incident radar electromagnetic waves back along the original direction, and have a big radar cross section. On the sea surface, the corner reflector or the corner reflector array can cover the real target and can produce obvious jamming, deception, and decoy to the off-position effect on the radar-guided weapon system [1, 2]. The inflatable icosahedral triangular corner reflector (ITCR) is commonly used for anti-ship missile defense [3–5]. Compared with other corner reflectors, icosahedral triangle corner reflectors are widely used. In this paper, compared with the octahedral triangle corner reflector, the simulation experiment of large-angle rotation in the terahertz frequency band is carried out for the ship model to study the performance of the icosahedral triangle corner reflector.

X. Li
School of Electronic Science, National University of Defense Science and Technology, Changsha, China

School of Physics, Electronics and Intelligent Manufacturing, Huaihua University, Huaihua, China

B. Deng (✉) · Q. Fu · H. Wang
School of Electronic Science, National University of Defense Science and Technology, Changsha, China
e-mail: dengbin_nudt@163.com

© The Author(s), under exclusive license to Springer Nature Switzerland AG 2024

S. Yeom (ed.), *4th International Conference on Electronics and Signal Processing*, Signals and Communication Technology,
https://doi.org/10.1007/978-3-031-36670-3_1

2 Imaging Algorithm

Without considering translation, inverse synthetic aperture radar (ISAR) imaging is a turntable imaging [6]. Considering the distance between the target (ships, corner reflectors) and the airborne radar, as well as the angle of radar scanning, the large-angle CBP fast imaging algorithm is used to image the ship and corner reflector on the water.

2.1 *Turntable Imaging*

As shown in Fig. 1a, the radar and the target are in the same plane. A uov coordinate system is established between the rotation centers of the radar and the target. The line connecting the rotation center of the radar and the target is the v axis. The line passing through the rotation center of the target and perpendicular to the v axis is the u axis, and the rotation center is the o point; the target takes point o as the rotation center to establish the xoy coordinate system. In the turntable imaging, the radar

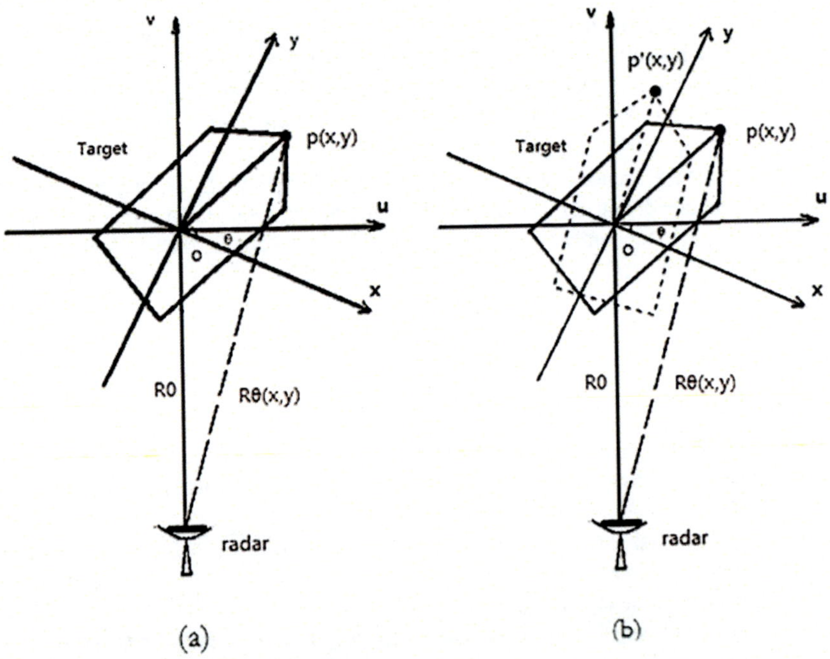

Fig. 1 Rotary table imaging geometric model

position remains unchanged, and the target rotates in the plane at a certain angular velocity with the rotation center o, as shown in Fig. 1b [7].

The included angle between xoy coordinate system and uov coordinate system is θ, and the transformation relationship between the two coordinate systems is:

$$\begin{cases} u = x\cos\theta + y\sin\theta \\ v = -x\sin\theta + y\cos\theta \end{cases} \tag{1}$$

or

$$\begin{cases} x = u\cos\theta - v\sin\theta \\ y = u\sin\theta + v\cos\theta \end{cases} \tag{2}$$

The distance from the point $p(x, y)$ on the target to the radar is:

$$\begin{aligned} R_\theta(x, y) &= \sqrt{(R_0 + v)^2 + u^2} \\ &= \sqrt{R_0^2 + x^2 + y^2 + 2R_0(y\cos\theta - x\sin\theta)} \end{aligned} \tag{3}$$

where R_0 is the distance from the radar to the target rotation center.

Suppose that the radar transmits a linear frequency-modulation signal, and the signal reflected from point $p(x, y)$ is mixed, the baseband echo signal obtained is:

$$S(k, \theta) = \int_y \int_x f(x, y) \exp\{-j2\pi k[R_\theta(x, y) - R_0]\} \mathrm{d}x\mathrm{d}y \tag{4}$$

where $f(x, y)$ is the distribution function of the electromagnetic scattering characteristics of the target, $k = 2(f_0 + \alpha t)/c$, f_0 is the carrier frequency of the frequency-modulation signal, α is the frequency modulation, c is the speed of light, and t is the fast time.

Under far-field conditions, there is [8]:

$$D^2 < \frac{R_0\lambda}{2} \tag{5}$$

where λ is the wavelength and D is the lateral expansion of the target. The electromagnetic wave meeting the far-field condition can be considered as a plane wave, and the distance from the target $p(x, y)$ to the radar can be equivalent to:

$$R_\theta(x, y) \approx R_0 + v = R_0 - x\sin\theta + y\cos\theta \tag{6}$$

The echo signals received by the radar at different observation positions are:

$$S(k, \theta) = \int_y \int_x f(x, y) \exp[-j2\pi k(-x\sin\theta + y\cos\theta)] \mathrm{d}x\mathrm{d}y \tag{7}$$

2.2 A CBP Fast Imaging Algorithm with Large Rotation Angle

Simply transform Eq. (8):

$$\begin{aligned} S(k,\theta) &= \int_y \int_x f(x,y)\exp\left[-j2\pi\left(-xk\sin\theta + yk\cos\theta\right)\right]\mathrm{d}x\mathrm{d}y \\ &= F\left(k_x, k_y\right) \end{aligned} \tag{8}$$

wherein $k_x = -x\sin\theta$ $k_y = k\cos\theta$, namely:

$$f(x,y) = \frac{1}{4\pi^2}\int_y\int_x F\left(k_x,k_y\right)\exp\left[j2\pi\left(xk_x + yk_y\right)\right]\mathrm{d}k_x\mathrm{d}k_y \tag{9}$$

Carry out Jacobian determinant coordinate transformation, and the above equation can be transformed into:

$$\begin{aligned} f(x,y) &= \frac{1}{4\pi^2}\int_{-\theta_m/2}^{\theta_m/2}\int_{k_{\min}}^{k_{\max}} kS(k,\theta)\exp\left[j2\pi k\left(-x\sin\theta + y\cos\theta\right)\right]\mathrm{d}k\mathrm{d}\theta \\ &= \frac{1}{2\pi}\int_{-\theta_m/2}^{\theta_m/2} G\left(-x\sin\theta + y\cos\theta\right)\mathrm{d}\theta \end{aligned} \tag{10}$$

where θ_m is the corner size,

$$G\left(-x\sin\theta + y\cos\theta\right) = \frac{1}{2\pi}\int_{k_{\min}}^{k_{\max}} kS(k,\theta)\exp\left[j2\pi k\left(-x\sin\theta + y\cos\theta\right)\right]\mathrm{d}k \tag{11}$$

This is the convolution back projection algorithm (CBP) [9]. Convolution back projection algorithm is divided into two steps: convolution and back projection. In convolution Eq. (11), the integral range of k is$[k_{\min}, k_{\max}]$, which does not meet the FFT condition, so it is necessary to frequency shift $k_{\min}$ in the actual algorithm, $k = k + k_{\min}$, that is:

$$\begin{aligned} G\left(-x\sin\theta + y\cos\theta\right) &= \frac{1}{2\pi}\int_0^B \left(k + k_{\min}\right) S\left(k + k_{\min},\theta\right) \\ &\exp\left[j2\pi\left(k + k_{\min}\right)\left(-x\sin\theta + y\cos\theta\right)\right]\mathrm{d}k \end{aligned} \tag{12}$$

where B is the bandwidth of spatial frequency. If the integral term is changed, the convolution term and the back projection integral term will become:

$$\begin{aligned} G\left(-x\sin\theta + y\cos\theta\right) &= \frac{1}{2\pi}\int_0^B \left(k + k_{\min}\right) S\left(k + k_{\min},\theta\right) \\ &\exp\left[j2\pi k\left(-x\sin\theta + y\cos\theta\right)\right]\mathrm{d}k \end{aligned} \tag{13}$$

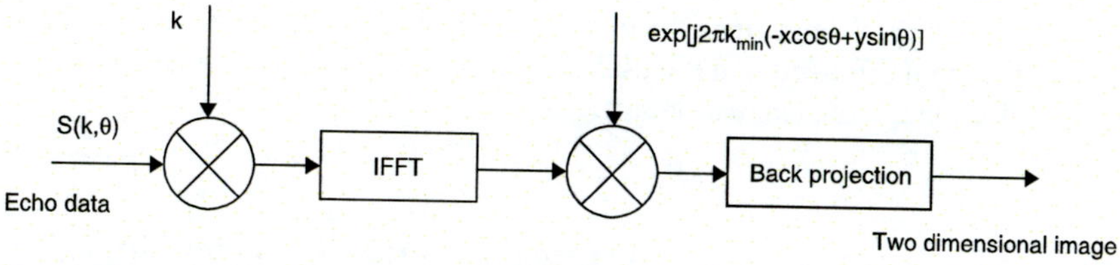

Fig. 2 Flowchart of convolution back projection algorithm

$$f(x, y) = \frac{1}{2\pi}\int_{-\theta_m/2}^{\theta_m/2} G(-x\sin\theta + y\cos\theta) \exp[j2\pi k_{\min}(-x\sin\theta + y\cos\theta)]\,\mathrm{d}\theta \tag{14}$$

The flowchart of the convolutional back projection algorithm is shown in Fig. 2.

In the back projection process, the integration range is $[-\theta_m/2, \theta_m/2]$, and all pixels need to be interpolated point by point and corner by corner in all visual angles. The interpolation process has a large amount of calculation and a long imaging time.

Assume that after convolution, the sampling point after FFT transformation is $u(\theta) = n \cdot \rho_r, n = 0, 1, 2, \cdots, N_r - 1$, where ρ_r is the range resolution and $u(\theta)$ is the interpolation point; the projection from the point (x, y) in the xoy coordinate to the viewing angle θ is $u'(\theta) = -x\sin\theta + y\cos\theta$. $u'(\theta)$ is the interpolated point; one-dimensional discrete interpolation is performed at interpolation point $u(\theta)$ to obtain the projection values of all pixels in view angle θ. The sampling values of each θ are listed in a row to form a one-dimensional discrete signal. The interpolation points correspond to the interpolated points. The imaging results of all pixels can be obtained by summing using the interpolation function interp1 [10–12].

Translate the data structure of the interpolation object, use one-dimensional interpolation, and the calculation process can traverse all points and angles. Compared with the conventional CBP algorithm, it does not need to make multiple cycles, and the interpolation efficiency is significantly improved.

3 Experiment

Altair Feko is a simulation software based on the method of moment (MOM) to realize full-wave electromagnetic field analysis of three-dimensional objects. Feko integrates multiple algorithms such as physical optical method (PO) to provide industry-leading hybrid solver, which can mix multiple frequency domain and time domain electromagnetic technologies to more effectively and accurately analyze electromagnetic scattering problems in complex media and electrically large sizes [13].

All experimental data are obtained through geometric modeling of ship, octahedral corner reflector, and icosahedral corner reflector and are simulated in FEKO electromagnetic simulation software.

3.1 Terahertz Radar Imaging of Corner Reflector Targets

The three-dimensional geometric modeling figure of octahedral corner reflector and icosahedral corner reflector is shown in Fig. 3. The octahedral corner reflector and icosahedral corner reflector are composed of multiple triangular corner reflectors. The faces of each triangular corner reflector are perpendicular to each other. Under the illumination of 220 GHz radar with elevation angle of 50° and azimuth angle of 45°, the image is shown in Fig. 4.

From Fig. 4, the icosahedral corner reflector has a more complex structure than the octahedral corner reflector. The octahedral corner reflector consists of a local scattering center. The icosahedral corner reflector has a local scattering center and

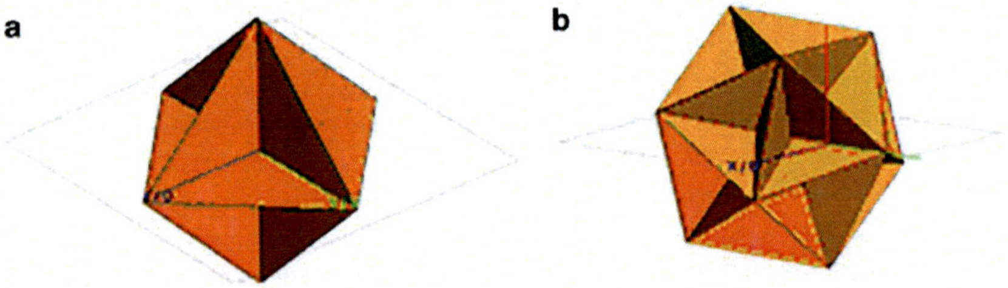

Fig. 3 Three-dimensional geometric model. (**a**) Octahedral corner reflector, (**b**) icosahedral corner reflector

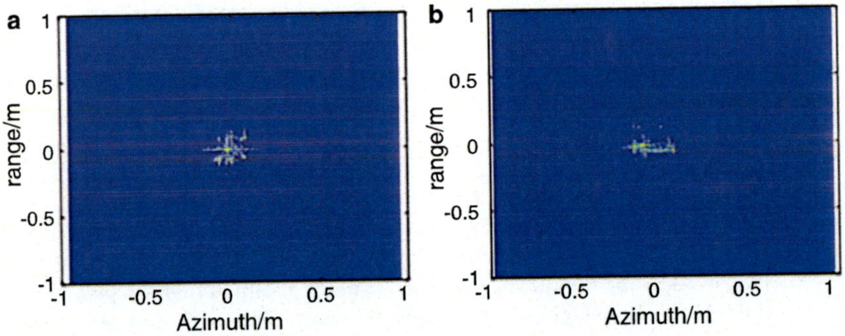

Fig. 4 220 GHz 2D imaging of octahedral corner reflector and icosahedral reflector. (**a**) Octahedral corner reflector, (**b**) icosahedral reflector

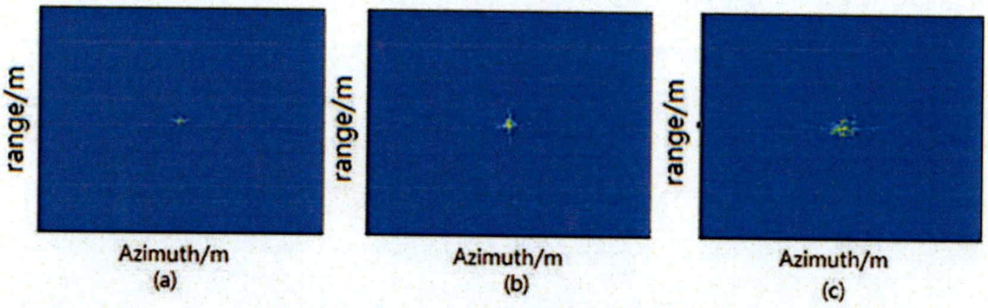

Fig. 5 Imaging of octagonal corner reflector. (**a**) 10 GHz, (**b**) 30 GHz, and (**c**) 220 GHz

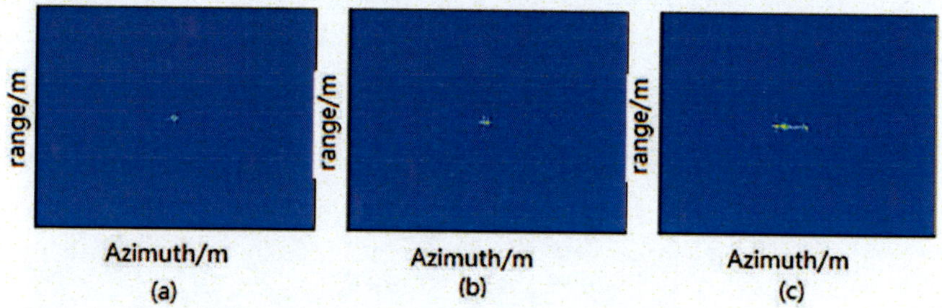

Fig. 6 Imaging of icosahedral corner reflector. (**a**) 10 GHz, (**b**) 30 GHz, and (**c**) 220 GHz

a spread scattering center. In the geometric model of the corner reflector, the sharp corners corresponding to the local scattering center can be roughly estimated.

Considering different electromagnetic bands, octahedral and icosahedral corner reflectors are imaged in X-band (10 GHz), Ka band (30 GHz), and terahertz band (220 GHz). The imaging of octahedral and icosahedral corner reflectors in different bands at the same elevation and azimuth is shown in Fig. 5. The imaging of icosahedral reflectors in different bands at the same elevation and azimuth is shown in Fig. 6.

From Figs. 5 and 6, the two-dimensional image of 10 GHz radar on octahedral corner reflector and icosahedral corner reflector is a strong scattering center, while the image of 220 GHz radar on octahedral corner reflector and icosahedral corner reflector is a strong scattering center and a spread scattering center. The frequencies of 10 GHz, 30 GHz, and 220 GHz radars increase in turn, and the shorter the wavelength, the clearer the imaging details of the target.

3.2 *THz Radar Imaging of Ship and Corner Reflector Targets*

The geometric model of the ship and the icosahedral corner reflector is shown in Fig. 7. The ship is 1.06 m long and 0.12 m wide. The ship is equipped with gun barrel and small transmitting antenna; the side length of the icosahedral reflector is 0.1 m,

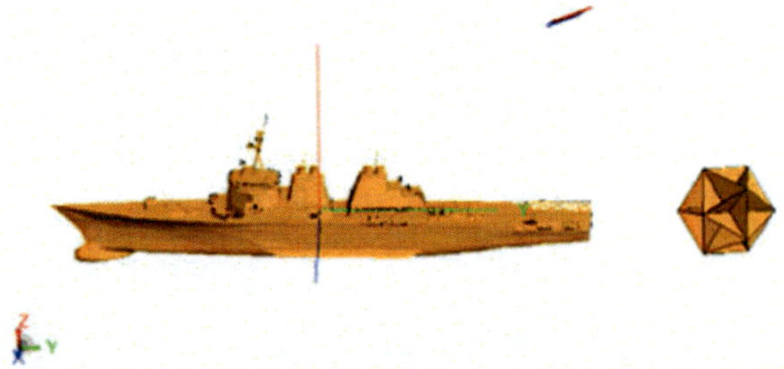

Fig. 7 Geometric model of ship and icosahedral corner reflector

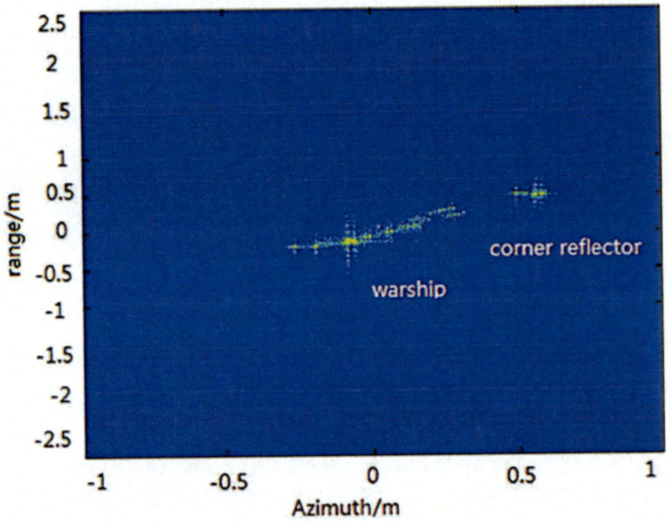

Fig. 8 220 GHz 2D imaging of ship and icosahedral corner reflector

and the imaging area of the ship and the icosahedral reflector is 1.92 × 0.48 m, 1.24 m wide horizontally. The elevation angle is 50° and the azimuth angle is 135°, which meet the requirements of imaging range resolution and azimuth resolution. The 220 GHz radar is used for imaging, and the imaging results are shown in Fig. 8.

From Fig. 8, it can be seen that the long ship shape can estimate the imaging position of the icosahedral corner reflector. Without prior knowledge, it is difficult to estimate the original geometry of the image.

In practice, multiple corner reflectors in the form of grouping interfere with or decoy the recognition of ships by guidance radars. The geometric model of the ship and four icosahedral corner reflectors is shown in Fig. 9. When the range resolution and azimuth resolution are satisfied, the 220 GHz radar two-dimensional imaging is carried out, and the imaging results are shown in Fig. 10.

From Fig. 10, the images of the four icosahedral corner reflectors have similar shapes.

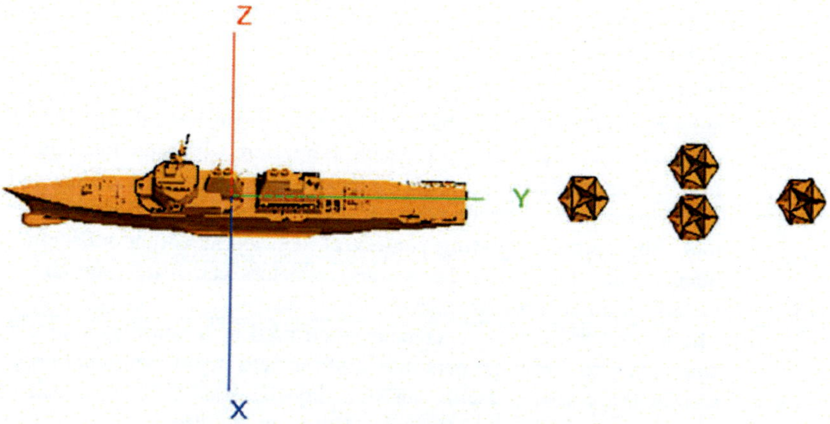

Fig. 9 Geometric model of ship and four icosahedral corner reflectors

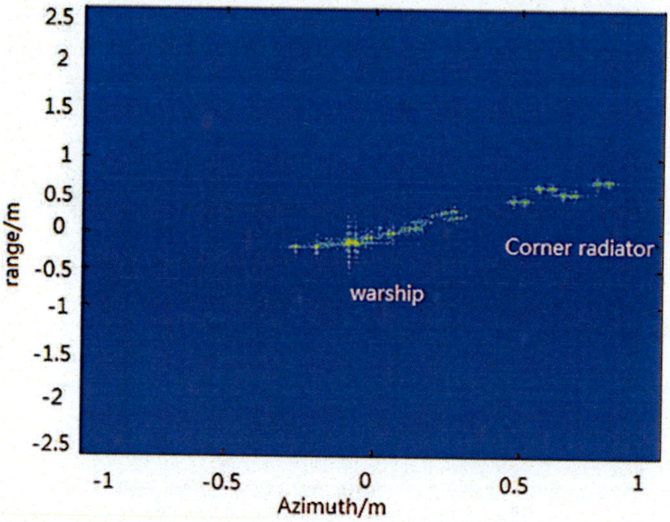

Fig. 10 220 GHz 2D imaging of ship and four icosahedral corner reflectors

4 Conclusion

Through 3D modeling of ships, octahedral corner reflectors, and icosahedral corner reflectors, imaging parameters are set on FEKO software, including plane wave, excitation source, polarization mode and incident direction, panel subdivision, and algorithm selection, and echo data of ships and corner reflectors are obtained. Through the large-angle CBP fast imaging algorithm, the two-dimensional imaging of ships and corner reflectors is obtained. Through the comparison of these two-dimensional images, it is concluded that:

(a) Compared with X-band and Ka band, the electromagnetic wavelength of terahertz band is shorter, which can better distinguish the details of targets, and it is more difficult for corner reflector to interfere with radar identification.
(b) In the terahertz frequency band, there are local scattering centers and spread scattering centers in the two-dimensional imaging of icosahedral corner reflectors. The spread scattering centers of octahedral corner reflectors are not obvious, and the icosahedral corner reflectors can better interfere with the identification of terahertz radars.
(c) The icosahedral corner reflector is often used as a jamming source for anti-ship guidance. From the perspective of far-field two-dimensional imaging, the icosahedral corner reflector can still play the role of jamming and inducing under the irradiation of THz radar. However, the density of corner reflectors laid on water, the types of interference sources such as corner reflectors, and the laying shape of interference source water surface will affect the recognition of targets on water by terahertz radar. There are many kinds of equipment on the surface of the ship. It is required that the interference source should be laid with different sizes of corner reflectors and aluminum strips according to the shape of the ship.

Acknowledgments This research was supported by the general program of the National Natural Science Foundation of China (61971427, 61871386, 62035014, 61921001, 62105363).

References

1. Zhiyuan, Z., JieQiu, Z., Shaobo, Q., & Hongliang, D. (2014). Research progress and Prospect of radar angle reflector. *Aviation Missile, 04*, 64–70.
2. Gang, C., Lu, Y., & Xiaoming, C. (2006). Tactical use of corner reflectors in the Navy. *Ship Electronic Countermeasures, 05*, 11–14.
3. Zhiyuan, Z., & Yuanyuan, Z. (2018). Analysis of electromagnetic scattering characteristics of a new icosahedral triangular corner reflector. *Command Control and Simulation, 40*(04), 133–137.
4. Kai, M., & Ma, W. (2020). Study on electromagnetic scattering characteristics of sea surface floating icosahedral reflector. *Digital Ocean and Underwater Attack and Defense, 3*(05), 437–442.
5. Li, C., Zhao, J., Yin, J., Zhang, G., & Shan, X. (2010). Analysis of RCS characteristic of dihedral corner and triangular trihedral corner reflectors. In *2010 5th International Conference on Computer Science & Education*, Hefei, China, pp. 40–43. https://doi.org/10.1109/ICCSE.2010.5593647
6. Peikang, H., Hongcheng, Y., & Xu, X. (2005). *Radar target characteristics*. Publishing House of Electronic Industry.
7. Yingbin, C., Bin, D., & Hongqiang, W. (2011). Turntable imaging algorithm based on projection slice theorem. *Signal Processing, 27*(09), 1380–1384.

8. Wang, B., Linxi, Z., Jiulun, F., & Weixin, X. (2009). A new filtering back projection imaging algorithm. *Computer Engineering and Applications, 45*(32), 16–18.
9. Xue, M., & Wang, Z. (1992). Application of back projection reconstruction method in microwave imaging. *Modern Radar, 06*, 74–79.
10. Mingmin, S., & Shaohui, Q. (2014). Research on ISAR fast CBP imaging algorithm. *Journal of Microwave, 30*(06), 31–35.
11. Jiang, W.-Q., Zhang, M., Nie, D., & Sun, R.-Q. (2016). Ship motion estimation from polarized Doppler spectra from ship wakes on two-dimensional sea surfaces. *Waves in Random & Complex Media, 26*(3), 284–300. https://doi.org/10.1080/17455030.2016.1139762
12. Zhang, M., & Li, J. (2019). Numerical simulation and analyses of SAR images from Ship Wakes. In *2019 International Applied Computational Electromagnetics Society Symposium (ACES)*, pp.1–2.
13. Junjuan, Z., Jingyuan, Y., & Chengfan, L. (2013). RCS simulation of artificial corner reflector based on FEKO platform. *Microelectronics and Computer, 30*(8), 79–81 + 85.

An Investigation of CNN-CARU for Image Captioning

Sio-Kei Im and Ka-Hou Chan

1 Introduction

Nowadays, media such as TV, website, news, and other media are usually connected with social communication. People are able to understand the information from the semantic feature through the given picture and video. Although it is published by the media without any description, people are also capable enough to interpret it by themselves without any description. On the contrary, it is very challenging for a machine to be able to make such abstract and fuzzy interpretations of it. Most applications of neural models focus on image segmentation, object recognition, and scene classification [17, 18]. Although these applications are very common and widely used in recent years, they are limited to image exploration and classification, which only provide information about class labels or identification of objects in an image, and rarely consider the properties of objects and the relationships between them [2].

In order to model the properties of images and the relationships between different objects, the people motion can be presented in different descriptions by different backgrounds [4]. It is used to locate where the object is, what it is doing, and much more of such activities. Image understanding requires recognition of objects, actions, and relationships [20]. For example, someone running in the background of a sports event means that people are running, but in the background of a disaster, it means that people are escaping [14, 19]. Another challenge is that the generated sentences must be syntactically and semantically correct. A good understanding of

S.-K. Im
Macao Polytechnic University, Macau, China

K.-H. Chan (✉)
Faculty of Applied Sciences, Macao Polytechnic University, Macau, China
e-mail: chankahou@mpu.edu.mo

© The Author(s), under exclusive license to Springer Nature Switzerland AG 2024

S. Yeom (ed.), *4th International Conference on Electronics and Signal Processing*, Signals and Communication Technology,
https://doi.org/10.1007/978-3-031-36670-3_2

a natural scene is based on the obtained features of natural language processing (NLP). This technology can be achieved through two main processing:

- First, a (pre-trained) convolutional neural network (CNN) is applied (a concatenate-designed classifier such as VGG and ResNet, etc. [8]), and the visual features of the image are encoded as a hidden state [34].
- Next, the language model usually consists of a recurrent neural network (RNN) that decodes the hidden states [21], for the target language through a word-by-word description of the image content to produce some tokens [33].

However, the captions generated are complete sentences that are generated as sequential words, depending on the properties of the image and the previously predicted words. If objects or scenes are misidentified at a particular time step in the sentence generation process, then sentences may be generated incorrectly [28].

2 Related Works

In practice, there are two approaches to image captioning in the image captioning literature: retrieval-based and generation-based image captioning [36]. Given an image, retrieval-based image captioning first searches the archive for relevant images and their descriptions and then assigns one (or more) of the most relevant pre-existing photo descriptions to the image caption. In fact, this approach can produce grammatically correct phrases if the archive is properly constructed, so it does not provide new captions from the current language [37]. In addition, generation-based approach makes the untrue assumption that there are always image-text combinations in the archive that are suitable for giving images [23]. After identifying the context of the image that is similar to the target image, they use a score of the decoded caption to compare the best caption from the list of produced captions [30]. The purpose of the current caption score is to measure the lexical similarity between the captions of the target images [31]. The target image was given the caption with the highest score among the captions of all retrieved photos. They show how their simple process outperforms some state-of-the-art image captioning techniques [1, 3].

For advanced use, [24] introduced a ranking mechanism as a framework for image feature encoding for captioning tasks. In [15] and [16], a method of Kernel Canonical Correlation Analysis was presented that links the largest training images to their caption and reduces images and text material to address a familiar region. In order to extract the best local feature to extract major information from the sequences images received, the SinP[N] [9] approach aims to compute the similarity scores between images and captions in the new generic domain [25], which will be computed for taking out the top-score features to describe the sequence of input images [35]. To minimize noise feature effects in the decoding process, [29] first used visually possible about the vision-based image retrieval, which produces the description of input images selected according to the order of the image caption.

From the retrieved image content, they calculated the word probabilities based on the current hidden state within RNN recurrent coding [22]. The word probability-based technique was used to convolve the current captions to retrieve the maximum score of the caption [38]. The previous principle established that given a caption, there is always a context that allows it. This evaluation is absolutely correct. Therefore, instead of using the selected sentences directly as descriptions of the images, the retrieved sentences are used again in the composition of the image caption. Furthermore, a number of studies have assumed that the labels of the images are determined by the labels that are retrieved directly from the word vocabulary. The target images are plotted in the meaning space using Markov models, the semantic distance between these images is derived using similarity measurements [27], and each existing sentence is parsed using the method in [6, 26]. The caption that is closest to the given image will be used as the caption for the most retrieved image. In the case of unknown image content, the generation of images is processed according to their global features in order to select an image group and thus derive a narrative. In order to generate phrases in combination with the retrieved images, a model for the learning of the meaning of predicate phrases is used [32]. Thus, the narrative is based on a selection of the most important phrases in general.

3 Proposed CNN-CARU for Image Captioning

Inspired by the above approaches, a CNN-CARU network model is introduced in this work. As presented in Fig. 1, the proposed captioning system consists of three stages: (1) The main stage is first to obtain the individual states produced from each level of concatenated-designed classifier [8] to achieve the feature extraction of CNN coding. Next (2) we calculate the scores of hidden state in the middle of each CNN layer, and (3) the one with highest score is selected and transferred to a stacked-designed CARU as the RNN decoding for the caption generating [5].

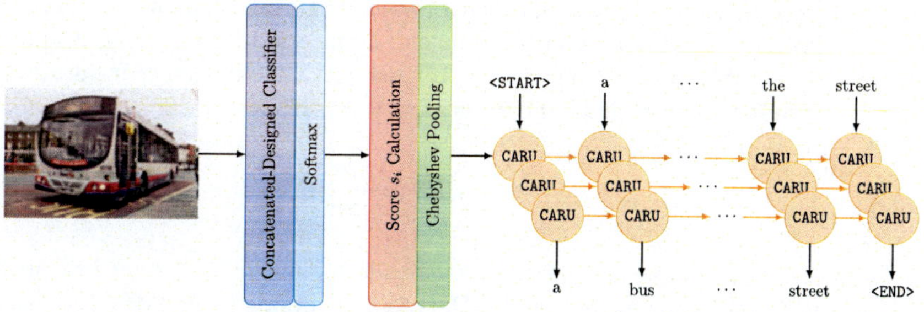

Fig. 1 The proposed CNN-CARU model will be used for the entire encoder–decoder architecture of the image captioning system. It demonstrates that each major stage is achieved through a specific approach

3.1 Concatenated-Designed Classifier for Feature Extracting

In general, a CNN classifier can usually be viewed as a tandem design architecture that implements filtering of images through a top-down stacking structure and provides the main feature products for further processing. A connected path is essentially a scalable, higher-level function extracting information from previous states. However, the size of the feature will be reduced, and the number of channels will increase as the feature passes through the CNN layer. Even though it captures the most important part of the classification target, the overall structure of the image is lost to a certain extent. Moreover, some useful information from the later layers may be discarded when information is extracted from the earlier layers. These problems become more obvious as the number of connections increases. This means that tandems are typically connected in up to five sets, and the final classifier will be pressured to make decisions. Comparatively, the former receives more original information, and the latter receives the extracted information. All classifiers independently make decisions based on their newly acquired features. Based on this idea, we propose to employ the number of t classifiers as

$$\forall t \bullet \sum \texttt{Classifier}_t\,(h_t) \equiv \mathbf{c}_t, \tag{1}$$

where the classification function `Classifier` transforms the last output feature h_t into a hidden vector $\mathbf{c}$.

3.2 Score Calculation for Determination

After each $\mathbf{c}_i$ calculating for $i = 0, \ldots, t$, these hidden vectors are designed to collect all outputs as input features to calculate the score of the CNNs state on the estimated target language that the decoder wants to predict. The purpose of this mechanism is the discovery of the relationship between the encoder and the decoder. This connection allows the decoder to capture global information to a certain extent, instead of having to infer from the current hidden state only. The determination is responsible for selecting which $\mathbf{c}_i$ is most relevant to the image feature and giving it high score when computing the context vector by

$$s_i = \frac{\exp\left(\mathbf{c}_i\right)}{\sum_k \exp\left(\mathbf{c}_k\right)}. \tag{2}$$

The higher the score, the more relative to the image (determined by $\arg\max\left(s_i\right)$) is for predicting the next word. Therefore, according to the `Softmax` function that ensures the sum of the elements of vector s_i is equal to one, the attention model produces the final context vector by weighting the encoder states h directly by

$$\mathbf{C} = \sum_{i}^{t} \mathbf{c}_i h_i. \tag{3}$$

However, we found that (2) uses a *Softmax* function, meaning that the scoring mechanism also inherited the disadvantages of *Softmax* during the training process: The convergence will be very slow, and the resulting context vectors will be very similar to each other, so that they will not be sufficiently discriminative. To alleviate this problem, we employ the advanced Chebyshev pooling technique [13] instead of maxpooling, and the results of convolution retain only local features and discard global information when pooling with decreasing resolution, making the decision after convolution and pooling more effective.

3.3 Multilayer CARU for Caption Generation

The encoder and decoder models perform the understanding and representation processing, respectively, and an additional extractor combines them to decide whether they should be included in the final caption. We use a recurrent neural network based on CARU [12] as the basis of our sequence selector. CARUs are recurrent units with multiple gate components and are expressed as follows:

$$x^{(t)} = \boldsymbol{w}_{vn} v^{(t)} + \boldsymbol{b}_{vn} \tag{4a}$$

$$n^{(t)} = \tanh\left(\left(\boldsymbol{w}_{hn} h^{(t)} + \boldsymbol{b}_{hn}\right) + x^{(t)}\right) \tag{4b}$$

$$z^{(t)} = \sigma\left(\boldsymbol{w}_{hz} h^{(t)} + \boldsymbol{b}_{hz} + \boldsymbol{w}_{vz} v^{(t)} + \boldsymbol{b}_{vz}\right) \tag{4c}$$

$$l^{(t)} = \sigma\left(x^{(t)}\right) \odot z^{(t)} \tag{4d}$$

$$h^{(t+1)} = \left(1 - l^{(t)}\right) \odot h^{(t)} + l^{(t)} \odot n^{(t)}, \tag{4e}$$

where the $\boldsymbol{w}$ and $\boldsymbol{b}$ are the training parameters (weight and bias) of linear layer. (4a) first receives the current input vector $v^{(t)}$ to $x^{(t)}$ as the input feature as the real-valued hidden-state vector at time step (t). This result would be used in the next hidden state and passed to the proposed content-adaptive gate. Next, (4b) just combines the parameters related to $h^{(t)}$ and $x^{(t)}$ to produce a new hidden state, and (4c) performs the update gate that is for hidden-state transitions. There is a Hadamard operator $\odot$ in (4d) that combines the update gate with the weights of the current features. This gate is named the content-adaptive gate, and it will affect the number of step-wise transitions instead of diluting the current hidden state. Finally, (4e) generates new hidden states for use in the next recurrent.

3.4 Implementation Detail

Our CNN-CARU uses VGGreNet [7] as the CNN encoder and a three-layer CARU network as the RNN decoder, with CARU feature input and output sizes of 256 and 1024, respectively, corresponding to the word embedding size [10, 11]. The output vocabulary for the generation of sentences was about 65k. In order to achieve the regularization of the image through the deep monitoring of the image embedding layer, we performed a random erasure of the image increments during the pre-processing. For this purpose, we also took advantage of the most frequently used words in the captions (used at least four times), which covered more than 95% of the occurrences of the words. For the learning strategy, with a batch size of 32 and a learning rate of 1e–3, we initially set all parameters to 10k iterations. In addition, there is a scheduler that is in control of the learning rate. It will reduce the rate of learning when the loss is becoming stagnant.

4 Conclusion

In this chapter, we provide a unique framework for image annotation by combining object detection and description-based descriptions for image description. A CNN-CARU architecture for detecting interesting objects is coupled with different descriptions. A consensus score compares the lexical similarity of the generated descriptions with the reference descriptions of the relevant photographs to provide a better description. We provide results comparing the proposed framework with a direct CNN-CARU architecture, and they demonstrate the potential of the proposed framework. Moreover, while reinforcement learning will soon be widely accepted for captioning in natural situations, the execution of image captions will undoubtedly be greatly improved by combining textual cues with visual information, which can overcome the same task for unsupervised learning models.

Acknowledgments This work is supported by the Macao Polytechnic University (Research Project RP/FCA-06/2023).

References

1. Al-Jamal, A. Z., Bani-Amer, M. J., & Aljawarneh, S. (2022). Image captioning techniques: A review. In *2022 International Conference on Engineering & MIS (ICEMIS)*. IEEE. https://doi.org/10.1109/icemis56295.2022.9914173
2. Aslam, A., & Curry, E. (2021). A survey on object detection for the internet of multimedia things (IoMT) using deep learning and event-based middleware: Approaches, challenges, and future directions. *Image and Vision Computing, 106*, 104095. https://doi.org/10.1016/j.imavis.2020.104095

3. Bai, S., & An, S. (2018). A survey on automatic image caption generation. *Neurocomputing, 311*, 291–304. https://doi.org/10.1016/j.neucom.2018.05.080
4. Beddiar, D. R., Nini, B., Sabokrou, M., & Hadid, A. (2020). Vision-based human activity recognition: A survey. *Multimedia Tools and Applications, 79*(41–42), 30509–30555. https://doi.org/10.1007/s11042-020-09004-3
5. Chan, K. H., & Im, S. K. (2022). Data stream classification by using stacked CARU networks. In *2022 IEEE International Conference on Big Data and Smart Computing (BigComp)*. IEEE. https://doi.org/10.1109/bigcomp54360.2022.00087
6. Chan, K. H., & Im, S. K. (2022). Partial attention modeling for sentiment analysis of big data. In *2022 7th International Conference on Frontiers of Signal Processing (ICFSP)*. IEEE. https://doi.org/10.1109/icfsp55781.2022.9924693
7. Chan, K. H., Im, S. K., & Ke, W. (2020). VGGreNet: A light-weight VGGNet with reused convolutional set. In *2020 IEEE/ACM 13th International Conference on Utility and Cloud Computing (UCC)*. IEEE . https://doi.org/10.1109/ucc48980.2020.00068
8. Chan, K. H., Im, S. K., & Ke, W. (2021). Multiple classifier for concatenate-designed neural network. *Neural Computing and Applications, 34*(2), 1359–1372. https://doi.org/10.1007/s00521-021-06462-0
9. Chan, K. H., Im, S. K., Ke, W., & Lei, N. L. (2018). SinP[N]: A fast convergence activation function for convolutional neural networks. In *2018 IEEE/ACM International Conference on Utility and Cloud Computing Companion (UCC Companion)*. IEEE. https://doi.org/10.1109/ucc-companion.2018.00082
10. Chan, K. H., Im, S. K., & Pau, G. (2022). Applying and optimizing NLP model with CARU. In *2022 8th International Conference on Advanced Computing and Communication Systems (ICACCS)*. IEEE. https://doi.org/10.1109/icaccs54159.2022.9785075
11. Chan, K. H., Im, S. K., & Zhang, Y. (2022). Optimization of language models by word computing. In *2022 The 6th International Conference on Graphics and Signal Processing (ICGSP)*. ACM. https://doi.org/10.1145/3561518.3561525
12. Chan, K. H., Ke, W., & Im, S. K. (2020). CARU: A content-adaptive recurrent unit for the transition of hidden state in NLP. In *Neural Information Processing* (pp. 693–703). Springer International Publishing. https://doi.org/10.1007/978-3-030-63830-6XXSlahUndXX58
13. Chan, K. H., Pau, G., & Im, S. K. (2021). Chebyshev pooling: An alternative layer for the pooling of CNNs-based classifier. In *2021 IEEE 4th International Conference on Computer and Communication Engineering Technology (CCET)*. IEEE. https://doi.org/10.1109/ccet52649.2021.9544405
14. Chen, N., Pan, X., Chen, R., Yang, L., Lin, Z., Ren, Y., Yuan, H., Guo, X., Huang, F., & Wang, W. (2021). Distributed attention for grounded image captioning. In *Proceedings of the 29th ACM International Conference on Multimedia*. ACM. https://doi.org/10.1145/3474085.3475354
15. Chen, Q., & Wang, Y. (2021). Key-performance-indicator-related state monitoring based on kernel canonical correlation analysis. *Control Engineering Practice, 107*, 104692. https://doi.org/10.1016/j.conengprac.2020.104692
16. Gao, X., Niu, S., & Sun, Q. (2019). Two-directional two-dimensional kernel canonical correlation analysis. *IEEE Signal Processing Letters, 26*(11), 1578–1582. https://doi.org/10.1109/lsp.2019.2939986
17. Gu, Y., Wang, Y., & Li, Y. (2019). A survey on deep learning-driven remote sensing image scene understanding: Scene classification, scene retrieval and scene-guided object detection. *Applied Sciences, 9*(10), 2110. https://doi.org/10.3390/app9102110
18. Hoeser, T., Bachofer, F., & Kuenzer, C. (2020). Object detection and image segmentation with deep learning on Earth observation data: A review—part II: Applications. *Remote Sensing, 12*(18), 3053. https://doi.org/10.3390/rs12183053
19. Hu, A., Chen, S., & Jin, Q. (2020). ICECAP: Information concentrated entity-aware image captioning. In *Proceedings of the 28th ACM International Conference on Multimedia*. ACM. https://doi.org/10.1145/3394171.3413576

20. Hu, M., Wang, H., Wang, X., Yang, J., & Wang, R. (2019). Video facial emotion recognition based on local enhanced motion history image and CNN-CTSLSTM networks. *Journal of Visual Communication and Image Representation, 59*, 176–185. https://doi.org/10.1016/j.jvcir.2018.12.039
21. Jiang, W., Ma, L., Jiang, Y.G., Liu, W., & Zhang, T. (2018). Recurrent fusion network for image captioning. In *Computer Vision – ECCV 2018* (pp. 510–526). Springer International Publishing. https://doi.org/10.1007/978-3-030-01216-8XXSlahUndXX31
22. Ke, W., & Chan, K. H. (2021). A multilayer CARU framework to obtain probability distribution for paragraph-based sentiment analysis. *Applied Sciences, 11*(23), 11344. https://doi.org/10.3390/app112311344
23. Li, R., Wang, Z., & Zhang, L. (2021). Image caption and medical report generation based on deep learning: a review and algorithm analysis. In *2021 International Conference on Computer Information Science and Artificial Intelligence (CISAI)*. IEEE. https://doi.org/10.1109/cisai54367.2021.00078
24. Liu, J., Cheng, K., Jin, H., & Wu, Z. (2022). An image captioning algorithm based on combination attention mechanism. *Electronics, 11*(9), 1397. https://doi.org/10.3390/electronics11091397
25. Mao, Y., Chen, L., Jiang, Z., Zhang, D., Zhang, Z., Shao, J., & Xiao, J. (2022). Rethinking the reference-based distinctive image captioning. In *Proceedings of the 30th ACM International Conference on Multimedia*. ACM. https://doi.org/10.1145/3503161.3548358
26. Peng, Y., Qi, J., & Yuan, Y. (2018). Modality-specific cross-modal similarity measurement with recurrent attention network. *IEEE Transactions on Image Processing, 27*(11), 5585–5599. https://doi.org/10.1109/tip.2018.2852503
27. Sattari, Z. F., Khotanlou, H., & Alighardash, E. (2022). Improving image captioning with local attention mechanism. In *2022 9th Iranian Joint Congress on Fuzzy and Intelligent Systems (CFIS)*. IEEE. https://doi.org/10.1109/cfis54774.2022.9756493
28. Sharma, H., Agrahari, M., Singh, S. K., Firoj, M., & Mishra, R. K. (2020). Image captioning: A comprehensive survey. In *2020 International Conference on Power Electronics & IoT Applications in Renewable Energy and its Control (PARC)*. IEEE. https://doi.org/10.1109/parc49193.2020.236619
29. Unar, S., Wang, X., Zhang, C., & Wang, C. (2019). Detected text-based image retrieval approach for textual images. *IET Image Processing, 13*(3), 515–521. https://doi.org/10.1049/iet-ipr.2018.5277
30. Wang, B., Li, G., Zhou, X., Chen, Z., Grossman, T., & Li, Y. (2021). Screen2words: Automatic mobile UI summarization with multimodal learning. In *The 34th Annual ACM Symposium on User Interface Software and Technology*. ACM. https://doi.org/10.1145/3472749.3474765
31. Wang, J., Xu, W., Wang, Q., & Chan, A.B. (2022). On distinctive image captioning via comparing and reweighting. *IEEE Transactions on Pattern Analysis and Machine Intelligence, 45*, 1–1. https://doi.org/10.1109/tpami.2022.3159811
32. Wang, L., Qian, X., Zhang, Y., Shen, J., & Cao, X. (2020). Enhancing sketch-based image retrieval by CNN semantic re-ranking. *IEEE Transactions on Cybernetics, 50*(7), 3330–3342. https://doi.org/10.1109/tcyb.2019.2894498
33. Wu, L., Xu, M., Wang, J., & Perry, S. (2020). Recall what you see continually using GridLSTM in image captioning. *IEEE Transactions on Multimedia, 22*(3), 808–818. https://doi.org/10.1109/tmm.2019.2931815
34. Xu, Y., Li, M., Cui, L., Huang, S., Wei, F., & Zhou, M. (2020). LayoutLM: Pre-training of text and layout for document image understanding. In *Proceedings of the 26th ACM SIGKDD International Conference on Knowledge Discovery & Data Mining*. ACM. https://doi.org/10.1145/3394486.3403172
35. Yan, C., Hao, Y., Li, L., Yin, J., Liu, A., Mao, Z., Chen, Z., & Gao, X. (2022). Task-adaptive attention for image captioning. *IEEE Transactions on Circuits and Systems for Video Technology, 32*(1), 43–51. https://doi.org/10.1109/tcsvt.2021.3067449

36. Yang, M., Liu, J., Shen, Y., Zhao, Z., Chen, X., Wu, Q., & Li, C. (2020). An ensemble of generation- and retrieval-based image captioning with dual generator generative adversarial network. *IEEE Transactions on Image Processing, 29,* 9627–9640. https://doi.org/10.1109/tip.2020.3028651
37. Zhao, D., Wang, A., & Russakovsky, O. (2021). Understanding and evaluating racial biases in image captioning. In *2021 IEEE/CVF International Conference on Computer Vision (ICCV).* IEEE. https://doi.org/10.1109/iccv48922.2021.01456
38. Zhu, H., Wang, R., & Zhang, X. (2021). Image captioning with dense fusion connection and improved stacked attention module. *Neural Processing Letters, 53*(2), 1101–1118. https://doi.org/10.1007/s11063-021-10431-y

Masked Face Recognition Using Transfer Learning Approaches

Mogeeb A. A. Mosleh, Abdulrahman Mohammed Al-Fakaih, Ayman Mohammed Al-Najar, Basheer Abdulraqeeb Farea, Mohammed Abdu Mugahed, Ahmed Salah Aldabe, Mohammed Abdallah Al-Kebsi, and Ayman AlSabry

1 Introduction

Artificial intelligence fields are growing up widely recently, attempting to simulate human capabilities and behaviors. Deep learning tools contributed widely to automating learning tasks for improving the efficiency of artificial intelligence techniques. Face recognition is an application area of AI, which involved both image processing and neural network techniques to recognize people. Face recognition techniques are a process of extracting selected biometric features from a human face to be used for identifying human identity. It is mostly used for security applications such as authentication purposes, tracking people, and criminal revelation. Research in face recognition achieved excellent progress with high performance and accuracy. However, there are several challenges and limitations that impacted the accuracy of face recognition techniques such as environmental, social, and external factors. For example, the accuracy of face recognition techniques is affected highly by environmental factors such as light intensity, angle of the face, and partial occlusion caused by wearable accessories. Some of these limitations caused by light and face angle are partially solved [1, 2] by using proper hardware requirements and other techniques. In addition, recognizing partial or occluded faces is still a challenge because of the absence of significant face features.

M. A. A. Mosleh (✉)
International University of Technology Twintech, Sana'a, Yemen

Software Engineering Department, Faculty of Engineering, Taiz University, Taiz, Yemen

A. M. Al-Fakaih · A. M. Al-Najar · B. A. Farea · M. A. Mugahed · A. S. Aldabe · M. A. Al-Kebsi · A. AlSabry
International University of Technology Twintech, Sana'a, Yemen

© The Author(s), under exclusive license to Springer Nature Switzerland AG 2024

S. Yeom (ed.), *4th International Conference on Electronics and Signal Processing*, Signals and Communication Technology,
https://doi.org/10.1007/978-3-031-36670-3_3

Challenges of face recognition techniques propagated during the COVID-19 epidemic [3], where governments impose on their citizens to wear medical masks to combat the spreading of the COVID-19 virus. Using masks that cover numerous details of the human face increases the difficulty of face recognition techniques by decreasing current system accuracy [4]. Additionally, the National Institute for Standards and Technology (NIST) reported that most face recognition algorithms after the COVID-19 pandemic show performance degradation when people's faces are masked.

In this paper, a pre-trained deep convolutional neural network model was developed to recognize masked faces by using a transfer learning approach. The proposed model was evaluated using a modified version of the CASIA [5] face dataset with synthetically generated masks. The proposed model is designed to solve the problem of recognition of individuals with masked faces using Inceptoin-ResNetV1. Our contribution includes (1) generating a faced mask database from the existing face image dataset, (2) selecting a suitable hyperparameter of Inceptoin-ResNetV1-based architecture to increase the learning rate and accuracy in identifying the masked face of the dataset, and (3) fine-tuning a pre-trained Inceptoin-ResNetV1 model on the modified dataset to achieve face recognition accuracy of 93%.

2 Methodology

2.1 Dataset

The CASIA dataset used for this study included more than 50,000 different individual face images. Then, a modified dataset was generated by applying Mask The Face techniques [6] on CASIA dataset to obtain a new dataset with synthetic face masks. Figure 1 illustrates the process of generating a synthetic face mask dataset workflow. Then, the generated dataset was divided into two sets including 70% for training and 30% for testing purposes.

In addition, a real masked face dataset that includes 100 images of 20 individuals with a mask and without a mask has been added to the modified dataset. The images were taken from various angles with good lighting and an indoor environment using a Canon 650D camera. The real masked face dataset was used as part of the testing dataset.

2.2 Deep CNN Architecture

A deep CNN model was developed here to achieve the study objective for recognition of masked faces as shown in Fig. 2, which included the following models:

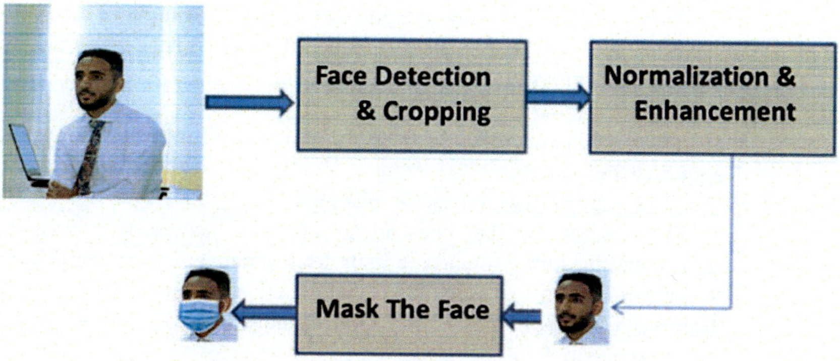

Fig. 1 Steps of generating a modified dataset with mask faces

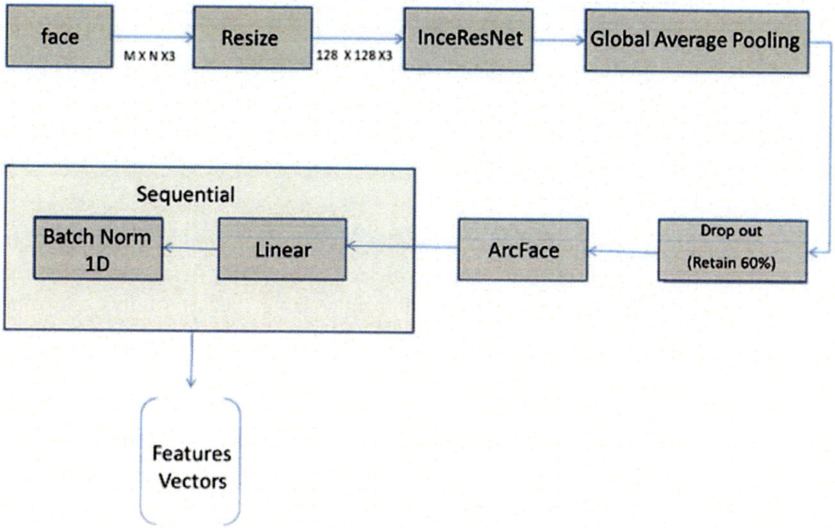

Fig. 2 Proposed deep CNN model architecture

- Resize: The real masked face dataset was resized to $128 \times 128 \times 3$ resolution to be suitable for input into the proposed model.
- Inception-ResNetV1:

  Transfer learning Inception-ResNetV1 (pre-trained on VGG2 dataset [7] deep CNN) was developed, which was inspired by both inception and the performance of a residual network. Inception-ResNet V1 was applied due to the limitation of hardware capability with current hyperparameters.
- Global average pooling

It was used as a pooling operation to replace the output of a fully connected layer in the CNN. It was used to reduce the input image size that fed into the fully connected layer.

- Dropout (retain 60%)

  It was used to decrease the overfitting issues by converting 40% of the dataset inputs to 0% randomly.

- Additive angular margin loss (ArcFace)

  The loss function has been adapted here to increase overall model performance and its ability to enforce high discriminative feature extraction. ArcFace is adapted here to be used as a loss function to obtain discriminative features from a masked face from the same individual [6].

- Linear sequential

  It was used to obtain 512 features as output from the ArcFace that result from 1792 features.

- Batch normalization

  It was applied to speed up the model learning time of learning by smoothing the optimization landscape significantly, which increases the model performance.

2.3 *Training Pipeline*

Finally, a pre-trained model was adapted in this study instead of starting from scratch by initialized weights to reduce the training loss convergence time since Inception-ResNetV1 was trained from a large dataset of faces. Global average pooling is used to reduce the computational cost of the model by reducing the output dimension of the convolutional layers. The sequential layers were adapted here to normalize and reduce the feature vector resulting from the ArcFace process.

3 Results and Discussion

3.1 *Training Phase*

In this phase, 70% of the dataset is used for training the CNN with face images and masked face images. The model shows a promising change in training loss after training it for 16 epochs with a batch size of 30. With these hyperparameters, the training loss starts to converge the dataset with an acceptable result. Using try-and-error experiments with different parameters, the model achieves the optimum results in epoch 51, as shown in Fig. 3. While with a higher number of epochs, the overfitting issues occur significantly.

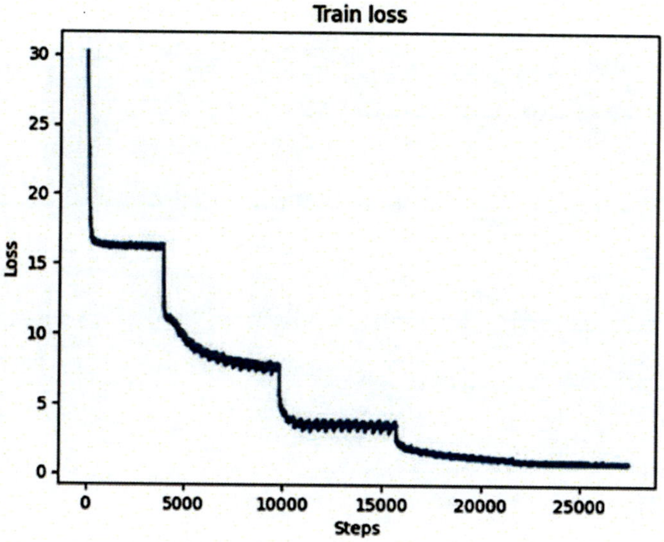

Fig. 3 Loss rate result in the training phase

3.2 Testing Phase

In this phase, the remaining 30% of the dataset is used to evaluate the model accuracy for recognition of the masked face images. Approximately, faces 7496 pairs of images for the same individual are used to evaluate the model as shown in Fig. 4. Different evaluation metrics are used in this phase such as confusion matrix, precision, F1 score, and recall to measure the model efficiency. The different metrics are involved because the confusion matrix is not considered to be a good measurement for accuracy where class imbalance datasets as the majority class overwhelm the errors in the minority class [8].

The testing phase has been conducted in the following steps for each pair of images: (1) Extract feature vectors of both images (individual face image and masked face image), (2) calculate and compare the extracted vectors using cosine distance algorithms, (3) both distances are verified by using a threshold value equal to 3, and (4) automatic matching on distance is performed to recognize the individual identity where the model results either negative if the distance is greater than the threshold value or positive in otherwise. Figures 5 and 6 illustrated the evaluation metrics results in the testing phase including the model accuracy, confusion matrix, precision, F1 score, and recall results.

Figure 5 showed that the confusion matrix indicated that the proposed model classified most of the testing dataset accurately for the same individual with less error rate. Moreover, Fig. 6 showed that the precision result was higher than the recall result, which indicated that the proposed model classified more positives

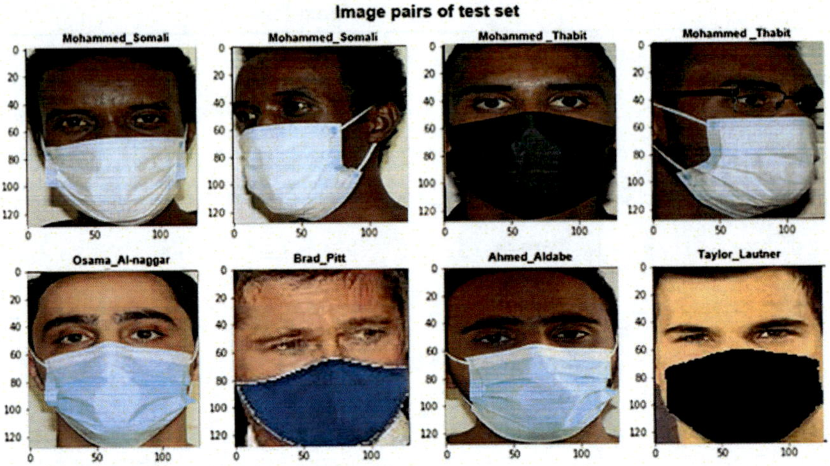

Fig. 4 Sample image pairs used in the testing phase

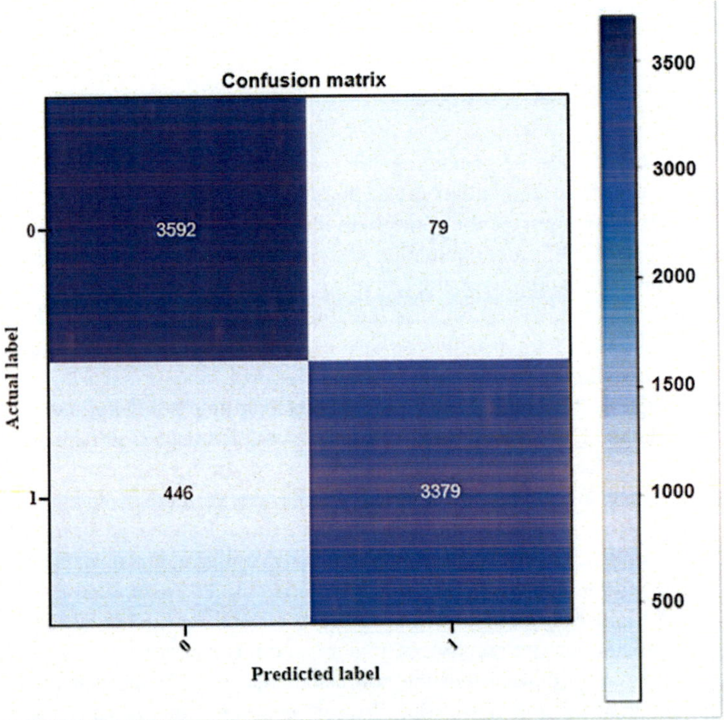

Fig. 5 Proposed model confusion matrix in testing phase

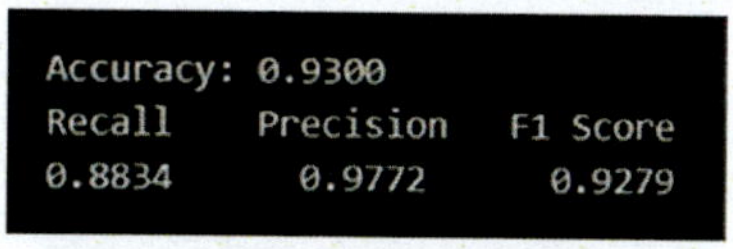

Fig. 6 Proposed model evaluation metrics results in testing phase

correctly. Finally, the drop in recall value is due to the imbalance of classes in the dataset. Additionally, the accuracy result in Fig. 6 showed that the accuracy result was 93%, which indicated that the proposed model was able to recognize the individual masked faces successfully with a high accuracy rate.

Results showed that the proposed model achieved 93% accuracy during the training phase when the tested pair contains masked faces. However, during the testing phase, the accuracy result of the model dropped to 85%, which indicates that the models do not fit to generalize all dataset masked faces. We assume that the problem is a result of the insufficient dataset in terms of class imbalance and the absence of a real masked face dataset. In general, with a comparison between the evaluation results with the art of works, some great improvements in face recognition techniques can be applied to identify individual masked faces by using transfer learning approaches.

4 Conclusion

The problem of occluded face recognition has gained researchers' interest, especially during the outbreak of coronavirus where individuals enforced to wear masks to reduce the infection spreading. Thus, wearable masks affect the reliability of existing face recognition models due to the absence of significant features of individual faces. To address this issue, this paper proposed a transfer learning-based approach by using a pre-trained DCNN model with a generated synthetic masked faces using the CASIA face dataset. The proposed model was developed by using mainly the Inception-ResNetV1 transfer learning model with some selected functions to achieve the study objective. The proposed model was evaluated using several metrics during both the training and testing phases. The proposed model result was able to recognize individual masked faces with a good accuracy rate of nearly 93%. Thus, the transfer learning model can be used to recognize individual identity even with masked faces, which can be used with some limitations for recognizing an individual with masked or covered face images.

5 Future Work

This work needs to extend the training with a wider range of pre-trained models to assess the performance and result of each transfer learning model. Furthermore, a large dataset needs to be considered that contains more than wearable masks with various skin colors that could improve the accuracy of results. Finally, the generalization model needs to be adapted well to obtain better masked faces so individuals can be set as a base task of future works.

6 Related Work

Recently, studies are contributed to make sense of developing several face recognition techniques. However, there are several limitations and challenges for most existing studies due to several reasons. These limitations and challenges result from natural, social, and dynamic changes in individuals [9]. Below are some research efforts related to the art of works.

Haibo Qiu et al. [10] proposed a method for face recognition that was robust to occlusions based on a single end-to-end deep neural network. They discovered that features from the DCNN were corrupted and tried to clean them with dynamically learned masks. They construct a dataset of occluded faces to train their model named Face Recognition with Occluded Mask (FROM). They reported that the experimental results on LFW dataset confirmed that FROM improved the accuracy of occluded faces and generalize well on normal faces.

Susanta Malakar et al. [11] conducted a reconstructive method to get partially reconstructed features of covered face parts. They adopted a deep learning method to recognize individual faces. First, they discarded the occluded part of the face and the remaining part of the face was applied by Principle Component Analysis (PCA). Then the principal components or the most significant eigenvector and its corresponding weights were calculated to reconstruct the occluded part. However, their proposed method could not reconstruct the occluded face part completely, but it could be used to improve the recognition accuracy by up to 15%.

Rubel Biswas et al. [12] identified a method for the present One-Shot Frequency-Dominant Neighborhood Structure (OSF-DNS), a new perceptual hashing method that showed advantages in two scenarios: (a) occluded face verification, i.e., matching occluded faces with their non-occluded versions, and (b) face classification. They compared the face verification performance of OSF-DNS with three perceptual hashing methods and with the features obtained from five deep learning techniques, using the occluded versions of six different facial datasets. The proposed method achieved accuracies between 69.24% and 99.46% depending on the dataset and always higher than the compared methods. For the face classification task, they compared the performance of OSF-DNS with the features obtained by four deep learning techniques. Experimental results on LFW and CFPW datasets showed that the proposed hashing method outperformed the results obtained with deep features with an accuracy of up to 89.53%.

Jiankang Deng et al. assumed (ArcFace) to earn highly discriminative features for face recognition. It has a clear geometric interpretation. They presented the most extensive experimental evaluation against all recent face recognition methods on ten face recognition benchmarks. They showed that ArcFace consistently outperforms and can be easily implemented with negligible computational overhead [13]. A real-world complex ISAR target recognition method has been suggested using deep adaptive learning on a limited platform by B Xue et al. It could handle real-world complex object recognition problems and leverage the relations among features, attributes, labels, and classes. Results demonstrated that the used method outperforms state-of-the-art methods [14]. Bin Xue et al. have presented active joint prior learning for real-world ISAR image blind. Extensive experimental results on ISAR image datasets showed the effectiveness of the proposed model for both synthesis and real-world noisy ISAR images, and the proposed method outperforms the state-of-the-art denoising methods [15]. A deep geometric learning approach is developed by using deep federated learning to improve the perception capabilities of convolutional neural networks and promote the representation maps' resolutions while achieving good recognition performance. They reported that InVision significantly outperforms the outstanding comparison approaches [16]. Ryo Hasegawa et al. tried to solve the problem of the effectiveness of road signs, changes in scale, and contrast for autonomous driving by learning road signs using a deep learning technique that is robust against scale changes. The used method showed higher accuracy in the detection and recognition of road signs than the faster region-based convolutional neural network and single-shot multibit detector methods [17]. A Discrete Wavelet Transform and Discrete Cosine Transform have been applied for face image recognition proposed by Samuel Lukas et al. The recognition process is performed by using a radial basis function neural network. Their results demonstrated that the best configuration for RBF was $8 \times 41 \times 41$ with a recognition rate of student faces of about 100% and 98% of the sample face images that were identified perfectly [18]. Haifeng Zhang and Shenjie Xu proposed an approach of weighted LTP to show facial features effectively. They divided the original face image into small blocks and calculated the pixels of the histogram and the LTP characteristic values. As a result of the experiment, they reported better performance on ORL and Yale Face Database [19]. Yordanka Karayaneva and Diana Hintea used the humanoid robot NAO to support object recognition of colors, shapes, typed words, and handwritten digits and operators. They included five neural network classifiers to apply for handwritten recognition of digits and operators. The five classifiers for handwritten recognition displayed highly accurate results that are within the range of 87–98% [20]. Yoshihiro Sato and Yue Bao conducted a three-dimensional face recognition approach. The three-dimensional shape of the face is acquired by using a Kinect sensor. Their experiment result showed that the authentication rate obtained by their method was 100% [21].

References

1. Boosting Masked Face Recognition with Multi-Task ArcFace [online].
2. Mosleh, M. A. A., et al. (2016). Ceph-X: Development and evaluation of 2D cephalometric system. *BMC Bioinformatics, 17*(19), 193–201.

3. Coronavirus Disease (COVID-19). Available online: https://www.cdc.gov/coronavirus/2019-ncov/prevent-getting-sick/index.html. Accessed on 2022.
4. Ngan, M., Grother, P., & Hanaoka, K. (2020). Ongoing Face Recognition Vendor Test (FRVT). Part 6A: Face recognition accuracy with masks using pre-COVID-19 algorithms [online].
5. Face Recognition: Biometric Authentication. Available online: https://www.nec.com/en/global/solutions/biometrics/face/. Accessed on 22 Aug 2022 [online].
6. Deng, J., et al. (2019). Arcface: Additive angular margin loss for deep face recognition. In *Proceedings of the IEEE/CVF Conference on Computer Vision and Pattern Recognition*.
7. Cao, Q., Shen, L., Xie, W., Parkhi, O. M., & Zisserman, A. (2018). VGGFace2: A dataset for recognizing faces across pose and age [online].
8. Baloch, B., Kumar, S., Rehman, A., & Syed, T. (2019). Focused Anchors Loss: Cost-sensitive learning of discriminative features for imbalanced classification [online].
9. Malek, S., Mosleh, M., Dhillon, S. K., & Milow, P. (2019). Bioimage informatics., pp. 993–1010.
10. Qiu, H., Gong, D., Li, Z., Liu, W., & Tao, D. (2021). End2End occluded face recognition by masking corrupted features. *IEEE Transactions on Pattern Analysis and Machine Intelligence, 44*, 6939–6952. [online].
11. Malakar, S., Chiracharit, W., & Chamnongthai, K. *2021 18th International Conference on Electrical Engineering/Electronics, Computer, Telecommunications and Information Technology (ECTI-CON)* [online].
12. Rubel Biswas, R., González-Castro, V., Fidalgo, E., & Alegre, E. (2021). A new perceptual hashing method for verification and identity classification of occluded faces. *Image and Vision Computing, 113*, 104245. [online].
13. Deng, J., Guo, J., Xue, N., & Zafeiriou, S. (2019). ArcFace: Additive angular margin loss for deep face recognition. *Proceedings of the IEEE Conference on Computer Vision and Pattern Recognition*, 4690–4699.
14. Xue, B., Yi, W., Jing, F., et al. (2021). Complex ISAR target recognition using deep adaptive learning. *Engineering Applications of Artificial Intelligence, 97*, 1–9.
15. Xue, B., He, Y., Jing, F., et al. (2021). Dynamic coarse-to-fine ISAR image blind denoising using active joint prior learning. *International Journal of Intelligent Systems, 36*(8), 4143–4166.
16. Xue, B., He, Y., Jing, F., et al. (2021). Robot target recognition using deep federated learning. *International Journal of Intelligent Systems, 36*(12), 7754–7769.
17. Hasegawa, R., Iwamoto, Y., & Chen, Y.-W. (2020). Robust Japanese road sign detection and recognition in complex scenes using convolutional neural networks. *Journal of Image and Graphics, 8*(3), 59–66. https://doi.org/10.18178/joig.8.3.59-66
18. Lukas, S., Mitra, A. R., Desanti, R. I., & Krisnadi, D. (2016). Implementing discrete wavelet and discrete Cosine transform with radial basis function neural network in facial image recognition. *Journal of Image and Graphics, 4*(1), 6–10. https://doi.org/10.18178/joig.4.1.6-10
19. Zhang, H., & Shenjie, X. (2016). The face recognition algorithms based on weighted LTP. *Journal of Image and Graphics, 4*(1), 11–14. https://doi.org/10.18178/joig.4.1.11-14
20. Karayaneva, Y., & Hintea, D. (2018). Object recognition in Python and MNIST dataset modification and recognition with five machine learning classifiers. *Journal of Image and Graphics, 6*(1), 10–20. https://doi.org/10.18178/joig.6.1.10-20
21. Sato, Y., & Bao, Y. (2018). 3D face recognition without using the positional relation of facial elements. *Journal of Image and Graphics, 6*(1), 33–38. https://doi.org/10.18178/joig.6.1.33-38

Part II
Signal Detection and Estimation

Analysis of Novel IpDFT Method for Frequency Estimation with Step-Changed Parameters

Tianchang Yang, Kai Wang, and Lu Tang

1 Introduction

In three-phase power system, it is very important to estimate the frequency phase and amplitude of balanced or unbalanced three-phase power system accurately and quickly. In the field of wireless communication, especially in frequency synthesizer, it is very important to monitor the output frequency accurately. The frequency estimation algorithm can be combined to achieve real-time detection when the output frequency changes, especially when the disturbance of environmental factors occurs. Therefore, frequency estimation algorithm is a very important and widely used algorithm, and there have been many algorithms to do a lot of research in this area, incorporating least squares method [1], linear autoregressive models [2], and Kalman filter [3]. In addition, several other methods, such as digital filter method [4], power spectrum method [5], and IpDFT method, can approximate frequency in the frequency domain. A unique IpDFT multi-DFT-bin approach for calculation of the frequency is proposed of a single tone signal with arguments jumping changes in Ref. [6]. However, the estimation of frequency in the unbalanced signal and the real-valued sinusoid signal is studied, with the elimination of the influence of jumping-changed parameters and elimination of the fence effect and spectrum leakage. This paper proposes an IpDFT approach for approximation of the frequency of a real-valued sinusoid signal with a step-changing amplitude/phase and the same frequency. Five distinct DFT bins are used for the requirement of the minimum the impact of the step-conversion arguments and spectrum leakage sufficiently, through which the simulation and testing results attain the objective of high accuracy of the frequency estimation.

T. Yang (✉) · K. Wang · L. Tang
Southeast University, Nanjing, China

© The Author(s), under exclusive license to Springer Nature Switzerland AG 2024

S. Yeom (ed.), *4th International Conference on Electronics and Signal Processing*, Signals and Communication Technology,
https://doi.org/10.1007/978-3-031-36670-3_4

However, these methods are suitable for frequency estimation of stable equilibrium signals without jump points, but not for the signals with unbalanced models and parameter jumps studied in this paper. Compared with the original IpDFT algorithm, the proposed algorithm takes more DFT bins, five DFT bins, which is according to the formula derivation in the following section. Although the formula derivation is more complex, the performance is better; especially when the parameter changes, the frequency estimation is more accurate than other algorithms.

2 Frequency Estimation Method

This paper applies to the case where a real-valued sinusoid signal with the step-changed amplitude and phase is generated, which can be described as

$$\begin{aligned} y(n) &= s(n) + w(n) \\ &= X_m \cos(\omega_0 n + \varphi_m) + w(n) \end{aligned} \tag{1}$$

where X_m and φ_m are the m-th ($m = 1, 2$) amplitude and phase of the unknown modulation signal; $\omega_0 = 2\pi f_0/f_s$ is the angular frequency, where f_0 is the signal frequency and f_s is the sampling frequency. $w(n)$ is the additive white Gaussian noise (AWGN) with variance σ_q^2.

The N-sample time sequence $y(n)$ where $n = 0, 1, \cdots, N$ is obtained by windowing $y(n)$ using a rectangular fenster of length N. We can rephrase $\omega_0 = 2\pi l_0/N = 2\pi(k_0 + \delta_0)/N$ as follows, where l_0 is the figure of captured sinusoid signal cycles and δ_0 and k_0 are the fractional and integer parts, respectively. The following expression is how the N-point DFT can be stated:

$$\begin{aligned} Y(k) &= S(k) + W(k) \\ &= \sum_{n=0}^{N-1} s(n) W_N^{nk} + W(k) \end{aligned} \tag{2}$$

where $S(k)$ is the DFT of the sequence $s(n)$, $W(k)$ is the DFT of the noise $w(n)$, and $W_N^k = e^{-j\left(\frac{2\pi}{N}\right)k}$. Let amplitude and phase step transition on P-th sample:

$$s(n) = \begin{cases} X_1 \cos(\omega_0 n + \varphi_1), & n = 0, 1, \cdots, P-1 \\ X_2 \cos(\omega_0 n + \varphi_2), & n = P+1, P+2, \cdots, N-1 \end{cases} \tag{3}$$

where X_m and φ_m ($m = 1, 2$) refer to any unknown amplitude and phase parameters. The DFT consequence at site k changes to:

$$
\begin{aligned}
2S(k) &= \sum_{n=0}^{N-1} X_1 \cos\left(\omega_0 n + \varphi_1\right) W_N^{nk} + \sum_{n=P}^{N-1} X_2 \cos\left(\omega_0 n + \varphi_2\right) W_N^{nk} \\
&= X_1 e^{j\varphi_1} \frac{1-\left(\lambda_0 W_N^k\right)^P}{1-\lambda_0 W_N^k} + X_1 e^{-j\varphi_1} \frac{1-\left(\lambda_0^* W_N^k\right)^P}{1-\lambda_0^* W_N^k} \\
&+ X_2 e^{j\varphi_2'} W_N^{kP} \frac{1-\left(\lambda_0 W_N^k\right)^{N-P}}{1-\lambda_0 W_N^k} + X_2 e^{-j\varphi_2'} W_N^{kP} \frac{1-\left(\lambda_0^* W_N^k\right)^{N-P}}{1-\lambda_0^* W_N^k}
\end{aligned} \tag{4}
$$

where $\lambda_0 = e^{j\omega_0}$ and X_1, X_2 are the amplitudes of signal before step change and after step change separately, φ_1 and φ_2 are the phases of signal before step-change and after step-change separately, P is the point where the signal has the step change, and N is the length of the signal. The index of the highest size of DFT can be used to do coarse frequency estimate, i.e., $\hat{k}_0 = \arg \max\limits_{k \in \{0,1,\ldots,N-1\}} (|S(k)|)$. Let's denote $\mu_1 = X_1 e^{j\varphi_1}$, $v_1 = X_1 e^{j\varphi_1} (\lambda_0)^P$, $\mu_2 = X_2 e^{j\varphi_2'}$, and $v_2 = X_2 e^{j\varphi_2'} (\lambda_0)^{N-P}$, where $\varphi_2' = \varphi_2 + P\omega_0$. We can rewrite Eq. (4) as:

$$
\begin{aligned}
2S(k) &= \frac{\mu_1 - v_1 W_N^{kP}}{1-\lambda_0 W_N^k} + \frac{\mu_1^* - v_1^* W_N^{kP}}{1-\lambda_0^* W_N^k} \\
&+ W_N^{kP} \frac{\mu_2 - v_2 W_N^{k(N-P)}}{1-\lambda_0 W_N^k} + W_N^{kP} \frac{\mu_2^* - v_2^* W_N^{k(N-P)}}{1-\lambda_0^* W_N^k} \\
&= \frac{(\mu_1+\mu_1^*)-(v_2+v_2^*)}{1-2\cos(\omega_0) W_N^k + \left(W_N^k\right)^2} - W_N^{kP} \frac{(v_1+v_1^*)-(\mu_2+\mu_2^*)}{1-2\cos(\omega_0) W_N^k + \left(W_N^k\right)^2} \\
&- W_N^k \frac{(\mu_1\lambda_0^*+\mu_1^*\lambda_0)-(v_2\lambda_0^*+v_2^*\lambda_0)}{1-2\cos(\omega_0) W_N^k + \left(W_N^k\right)^2} + W_N^{k(P+1)} \frac{(v_1\lambda_0^*+v_1^*\lambda_0)-(\mu_2\lambda_0^*+\mu_2^*\lambda_0)}{1-2\cos(\omega_0) W_N^k + \left(W_N^k\right)^2}
\end{aligned} \tag{5}
$$

Let $a_1 = (\mu_1 + \mu_1^*) - (v_2 + v_2^*)$, $a_2 = (v_1 + v_1^*) - (\mu_2 + \mu_2^*)$, $a_3 = (\mu_1\lambda_0^* + \mu_1^*\lambda_0) - (v_2\lambda_0^* + v_2^*\lambda_0)$, $a_4 = (v_1\lambda_0^* + v_1^*\lambda_0) - (\mu_2\lambda_0^* + \mu_2^*\lambda_0)$ and $b = \cos(\omega_0)$. Then we can get a simpler equation:

$$
2S(k)\left[1 + \left(W_N^k\right)^2\right] = a_1 - W_N^{kP} a_2 - W_N^k a_3 + W_N^{k(P+1)} a_4 + 4bS(k) W_N^k \tag{6}
$$

As we have five unknown numbers a_1, a_2, a_3, a_4, b, when any five different DFT bins $S(k_i)$ ($i = 1, 2, \cdots, 5$) are obtained, it will be enough to help the algorithm get the accurate result and without adding excessive computational burden. So the following linear equations according to Eq. (6) can be shown:

$$
\mathbf{S}_k = \mathbf{W}_k \boldsymbol{\eta} \tag{7}
$$

where $\mathbf{S}_k = \left[\left(1+\left(W_N^{k_1}\right)^2\right) S(k_1), \left(1+\left(W_N^{k_2}\right)^2\right) S(k_2), \cdots, \left(1+\left(W_N^{k_5}\right)^2\right) S(k_5)\right]^T$, $\boldsymbol{\eta} = [a_1, a_2, a_3, a_4, b]$ and

$$\mathbf{W}_k = \begin{bmatrix} 1 - W_N^{k_1 P} & -W_N^{k_1} & W_N^{k_1(P+1)} & 2W_N^{k_1} S(k_1) \\ 1 - W_N^{k_2 P} & -W_N^{k_2} & W_N^{k_2(P+1)} & 2W_N^{k_2} S(k_2) \\ 1 - W_N^{k_3 P} & -W_N^{k_3} & W_N^{k_3(P+1)} & 2W_N^{k_3} S(k_3) \\ 1 - W_N^{k_4 P} & -W_N^{k_4} & W_N^{k_4(P+1)} & 2W_N^{k_4} S(k_4) \\ 1 - W_N^{k_5 P} & -W_N^{k_5} & W_N^{k_5(P+1)} & 2W_N^{k_5} S(k_5) \end{bmatrix}$$

According to Eq. (7), $\boldsymbol{\eta}$ can be estimated in the noisy condition as:

$$\hat{\boldsymbol{\eta}} = \tilde{\mathbf{W}}_k^{-1} \mathbf{X}_k \tag{8}$$

where $\tilde{\mathbf{W}}_k$ and $\mathbf{X}_k$ are the observed values of $\mathbf{W}_k$ and $\mathbf{S}_k$ in the noisy condition. $\hat{\boldsymbol{\eta}}$ is the estimated value of $\boldsymbol{\eta}$. According to $b = \cos(\omega_0)$, the angular frequency ω_0 can be estimated as:

$$\hat{\omega}_0 = \cos^{-1}\left(\ln \hat{b}\right) \tag{9}$$

When ω_0 has been estimated, we can get the following equation with any four different DFT bins $X(k_i)$, $i = 1, 2, 3, 4$ according to Eq. (5):

$$\mathbf{W}_{Aest} \begin{bmatrix} \hat{\upsilon}_1 \\ \hat{\upsilon}_2 \\ \hat{u}_1 \\ \hat{u}_2 \end{bmatrix} = \begin{bmatrix} 2\left(1 - W_1^1\right)\left(1 - W_2^1\right) X(k_1) \\ 2\left(1 - W_1^2\right)\left(1 - W_2^2\right) X(k_2) \\ 2\left(1 - W_1^3\right)\left(1 - W_2^3\right) X(k_3) \\ 2\left(1 - W_1^4\right)\left(1 - W_2^4\right) X(k_4) \end{bmatrix} \tag{10}$$

where $W_1^i = \lambda_0 W_N^{k_i}$ $(i = 1, 2, 3, 4)$, $W_2^i = \lambda_0{}^* W_N^{k_i}$ $(i = 1, 2, 3, 4)$ and

$$\mathbf{W}_{Aest} = \begin{bmatrix} \left(1 - W_1^1\right)^P \left(1 - W_2^1\right) & \left(1 - W_2^1\right)^P \left(1 - W_1^1\right) & W_N^{k_1 P} \left(1 - W_1^1\right)^{N-P} \left(1 - W_2^1\right) & W_N^{k_1 P} \left(1 - W_2^1\right)^{N-P} \left(1 - W_1^1\right) \\ \left(1 - W_1^2\right)^P \left(1 - W_2^2\right) & \left(1 - W_2^2\right)^P \left(1 - W_1^2\right) & W_N^{k_2 P} \left(1 - W_1^2\right)^{N-P} \left(1 - W_2^2\right) & W_N^{k_2 P} \left(1 - W_2^2\right)^{N-P} \left(1 - W_1^2\right) \\ \left(1 - W_1^3\right)^P \left(1 - W_2^3\right) & \left(1 - W_2^3\right)^P \left(1 - W_1^3\right) & W_N^{k_3 P} \left(1 - W_1^3\right)^{N-P} \left(1 - W_2^3\right) & W_N^{k_3 P} \left(1 - W_2^3\right)^{N-P} \left(1 - W_1^3\right) \\ \left(1 - W_1^4\right)^P \left(1 - W_2^4\right) & \left(1 - W_2^4\right)^P \left(1 - W_1^4\right) & W_N^{k_4 P} \left(1 - W_1^4\right)^{N-P} \left(1 - W_2^4\right) & W_N^{k_4 P} \left(1 - W_2^4\right)^{N-P} \left(1 - W_1^4\right) \end{bmatrix}$$

Then the amplitude and phase can be calculated in view of the relationship as $\mu_1 = A_1 e^{j\varphi_1}$, $\upsilon_1 = A_1 e^{j\varphi_1} (\lambda_0)^P$, $\mu_2 = A_2 e^{j\varphi_2'}$, and $\upsilon_2 = A_2 e^{j\varphi_2'} (\lambda_0)^{N-P}$, so A_1, A_2, φ_1, and φ_2 can be estimated as:

$$\begin{cases} \hat{A}_1 = \mathrm{abs}\left(\hat{\mu}_1\right) \\ \hat{A}_2 = \mathrm{abs}\left(\hat{\mu}_2\right) \\ \hat{\varphi}_1 = \arg\left(\hat{\mu}_1\right) \\ \hat{\varphi}_2 = \arg\left(\hat{\mu}_2\right) - P\hat{\omega}_0 \end{cases} \tag{11}$$

As previously stated, the proposed approach can use any five distinct DFT bins to estimate the frequency. For the best estimation accuracy, the DFT bin with the greatest magnitude $Y(k_0)$, $Y(k_0 \pm 1)$, and $Y(k_0 \pm 2)$ should be generally recommended.

3 The Performance of Raised Algorithm

In this section, the verification of the theoretical derivation of the previous chapter is presented by simulation. For all experiments in this section, the sampling points set $N = 128$. Limited to typesetting restrictions, we will only take $A_1 = 0.9, A_2 = 1$ and $\varphi_1 = 10\pi/180$, $\varphi_2 = 80\pi/180$as examples for signals with symbol transition in the following analysis. The DFT bin with the biggest amplitude $Y(k_0)$ and its surrounding DFT bins $Y(k_0 \pm 1)$ and $Y(k_0 \pm 2)$ are generally recommended in practical application for the optimal estimation accuracy. Based on the estimated outcomes, we compare the manifestation of each algorithm. For all tests, we set $N = 128$, amplitudes of two signals are 0.9 and 1, and phases of two signals are selected randomly. The properties of our algorithm are comparative with those of the XJX method in Ref. [6], the UFE method in Ref. [3], the BFEL method in Ref. [3], the IpDFT method in Ref. [7] and IpN3DFT method in Ref. [8]. To examine the statistical qualities of every argument, 2000 runs are carried out. The definition of the mean square error (MSE) for the degree of exactitude of the calculating algorithm estimation is:

$$\text{MSE}\,(\omega_0) = 10\,\text{lg}\left[\frac{1}{M}\sum_{i=1}^{M}\left(\hat{\omega}_0(i) - \omega_0\right)^2\right] \tag{12}$$

Figure 1 portrays the relation between l_0 and MSEs of different frequency estimation algorithms when SNR = 40 dB and $P = 10$. The UFE method and IpDFT method are no longer effective due to the interference of side lobes and the spectrum leakage raised by negative frequency, for the signal model is the real-valued sinusoid signal with step-change amplitude and phase. On the other hand, IpN3DFT method is a method without the consideration of the step-change parameters, which can only accurately estimate the frequency the real-valued sinusoid signal. The method proposed in this paper takes into consideration not only the interference of side lobes and the spectrum leakage raised by negative frequency component but also the step-change parameters that have an impact on the DFT bins. Therefore, the proposed algorithm can achieve frequency approximation of the real-valued sinusoid signal with high precision. As shown in Fig. 1, with the change of l_0, the performance of our algorithm remains stable and optimal, where our mean reaches the lowermost MSE, around −80 dB.

When $l_0 = 1.48$ and $P = 10$, Fig. 2 demonstrates the connections between SNR and MSEs of several techniques. The method's MSEs reduce linearly as SNR rises, implying that these approaches can accurately estimate frequency when amplitude

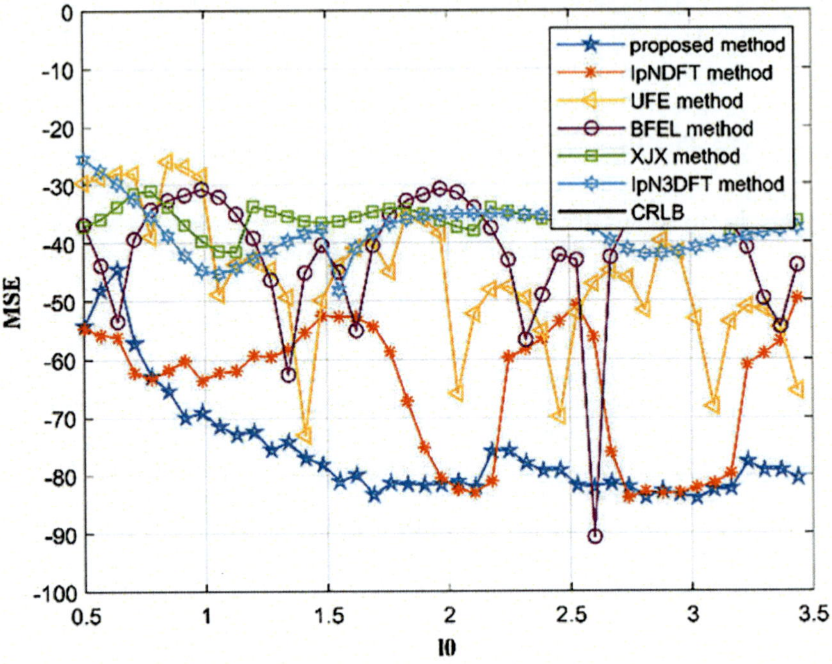

Fig. 1 MSE of $\hat{\omega}_0$ versus l_0 with SNR = 40 dB, $P = 64$

and phase step transitions on the P-th sample. The results suggest that our technique performs better at different SNR levels.

Figure 3 describes the relationships between P and MSEs of different algorithms when $l_0 = 1.48$ and SNR = 40 dB. When $P = 0$ or $P = N - 1$, the signal deviates into a real-value sinusoid signal in fact, and our means ultimately deviates into an IpN3DFT method in Ref. [8]. Let P changes from 1 to 64. It could be found that the performance of our algorithm remains stable and does not change with the change of P, where the minimum MSE appears in our algorithm, around −80 dB.

According to the above simulation results, select a group of experimental conditions for error comparison ($l_0 = 1.48$, SNR = 40 dB, $P = 64$), and the results shown in the following Table 1 can be obtained.

4 Consequence Under the Condition of Signal with Parameter Jump

Simulation findings for step-transition signals in magnitude and phase with SNR = 40 dB are shown in Fig. 4a, b. The magnitude and phase step sizes are

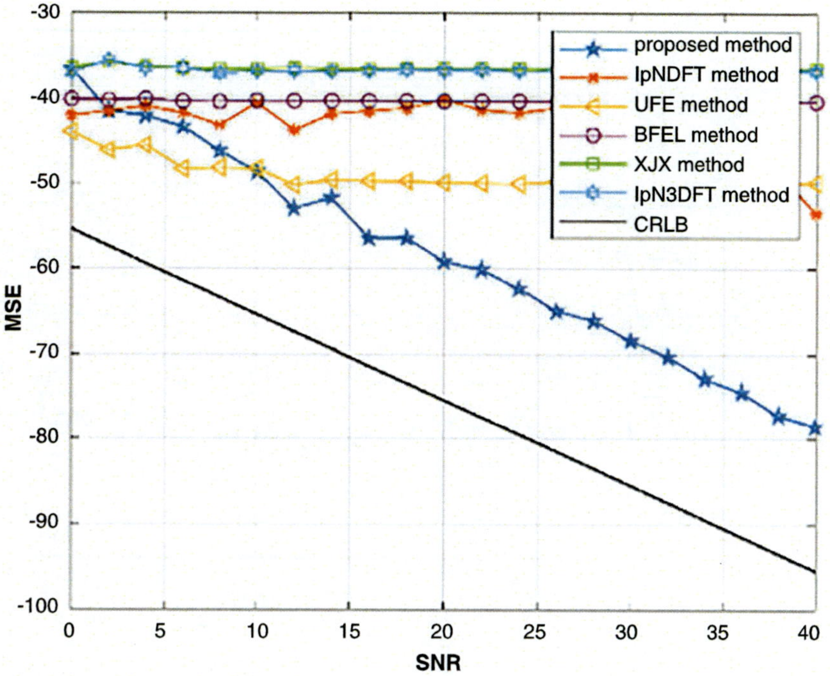

Fig. 2 MSE of $\hat{\omega}_0$ versus SNR with $l_0 = 1.48$, $P = 64$

both set at 0.1. Step changes $\pi/18$ take place every $t = 0.15$ s and are released every $t = 0.3$ s. When the step change happens and is released, our method's estimation can coincide with the reference frequency, while the other methods are very oscillating because they do not take the effects of the picket fence effect and the spectrum leakage into account. During the whole step-change interval, our algorithm's MSE remains stable and can cursorily get to -70 dB. As the influence of step-change parameters is not considered, the estimation result of IpN3DFT method oscillates within the step-change parameter interval, but the estimation performance is stable and accurate outside the step-change parameter interval. Side-lobe interference and spectrum leakage caused by the negative frequency component impair the estimation results of the single-tone signal model, including UFE method, XJX method, BFEL, and IpDFT methods. Therefore, whether within or outside the step-change parameter interval, the results are expected to be very bad.

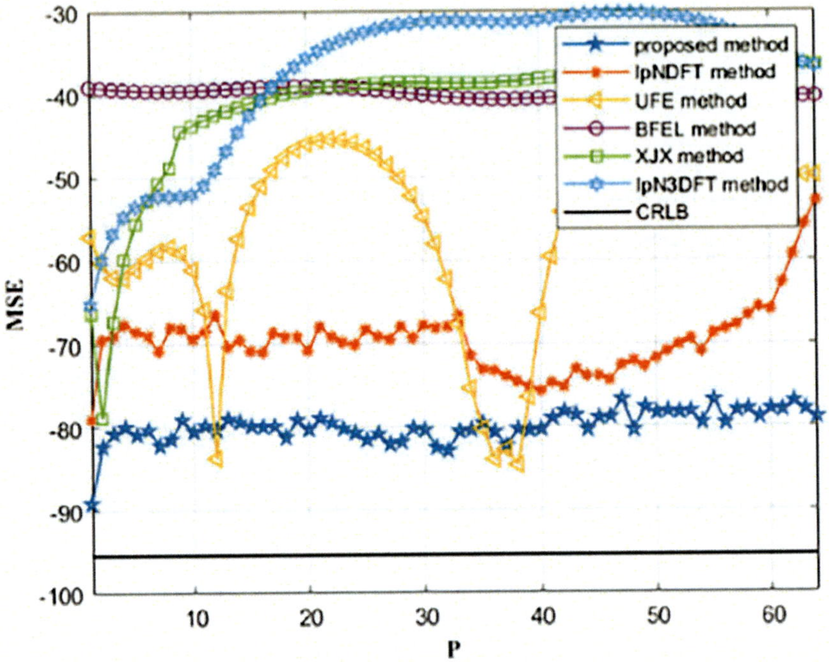

Fig. 3 MSE of $\hat{\omega}_0$ versus P with $l_0 = 1.48$, SNR = 40 dB

Table 1 The MSE of each algorithm when $l_0 = 1.48$, SNR = 40 dB, $P = 64$

Algorithm	Proposed method	IpNDFT	UFE	BFEL	XJX	IpN3DFT
MSE(dB)	−79.95	−52.13	−50.34	−40.18	−37.45	−37.42

It can be clearly seen from the figure that the algorithm proposed by us, in this system, can accurately estimate the output frequency value without fluctuation caused by step changing.

5 Conclusion

This paper proposes a new IpDFT method and theoretical variance calculation formula, which can deal with the estimation errors caused by parameter step changing more effectively. The minimum MSE of our proposed algorithm can reach about −80 dB, which is much better than other algorithms. This work will be studied in the direction of higher frequency in the future, hoping to combine some more practical systems, such as the phase-locked loop (PLL) frequency synthesizer system.

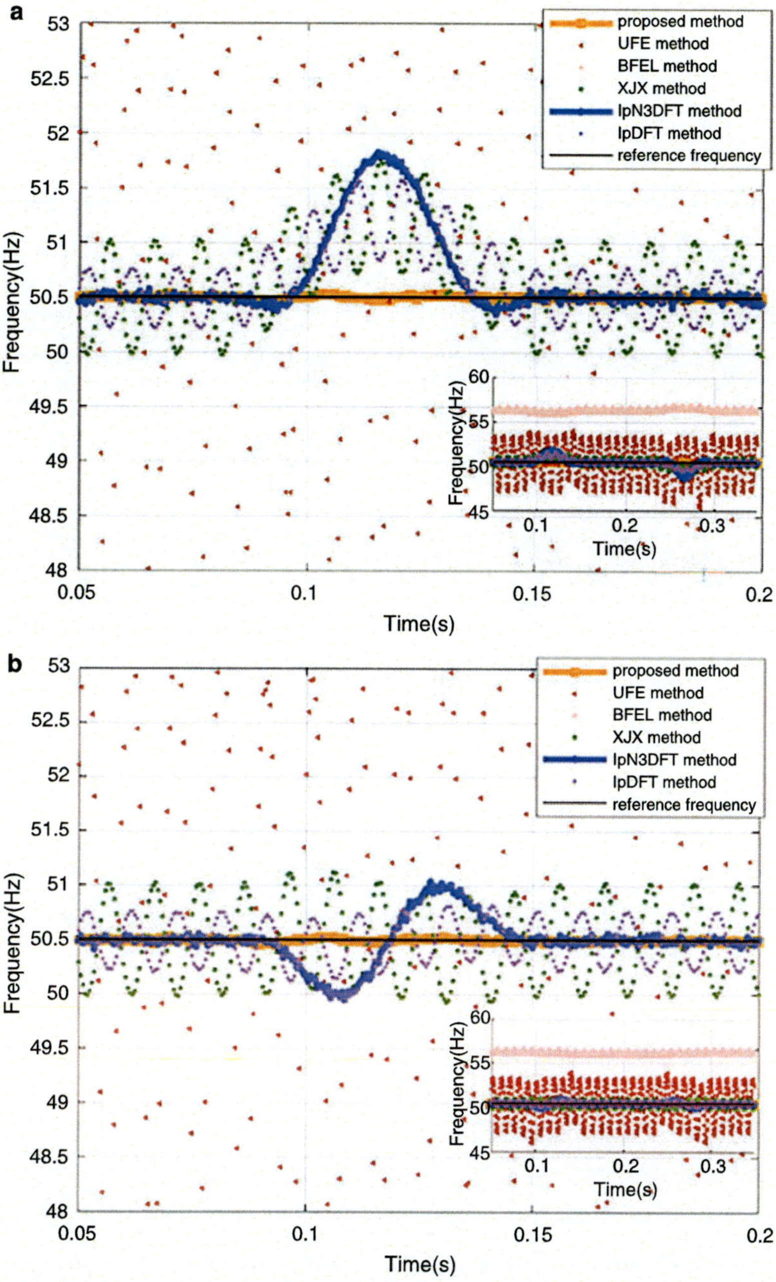

Fig. 4 Estimated frequency of emulation for step-transition signal in the (**a**) magnitude (**b**) phase when SNR=40 dB

References

1. Han, J., Li, W., Xiao, J., Fengand, J., Yang, Q., & Yu, S. (2015). Frequency offset estimation with multi-steps interpolation for coherent optical systems. *IEEE Photonics Technology Letters, 27*(19), 2011–2014.
2. Aboutanios, E., & Mulgrew, B. (2005). Iterative frequency estimation by interpolation on fourier coefficients. *IEEE Transaction on Signal Processing, 53*(4), 1237–1242.
3. Xiao, J. (2015). Low complexity FFT-based frequency offset estimation for MQAM coherent optical systems. *IEEE Photonics Technology Letters, 27*, 1371–1374.
4. Baradaraniand, F., Zadeh, M. R. D., & Zamani, M. A. (2015). A phase-angle estimation method for synchronization of grid-connected power-electronic converters. *IEEE Transactions on Power Delivery, 30*(2), 827–835.
5. Ren, J., & Kezunovic, M. (2011). Real-time power system frequency and phasors estimation using recursive wavelet transform. *IEEE Transactions on Power Delivery, 26*(3), 1392–1402.
6. Wang, K., & Liu, S. (2022). *Novel interpolation method of multi-DFT-bins for frequency estimation of signal with parameter step change* (Vol. 71, p. 1). IEEE Transactions on Instrumentation and Measurement.
7. Peng, A., Ou, G., & Shi, M. (2014). Frequency estimation of single tone signals with bit transition. *IET Signal Processing, 8*(9), 1025–1031.
8. Wang, K., Wenand, H., & Li, G. (2021). Accurate frequency estimation by using three-point interpolated discrete fourier transform based on rectangular window. *IEEE Transactions on Industrial Informatics, 17*(1), 73–81.

Joint Angle-Polarization-Range Estimation for EVS-FDA-MIMO Radar

Tiantian Zhong and Haihong Tao

1 Introduction

Target parameter estimation is an enduring signal processing problem that has always raised persistent attention within the radar scientific community. The problem of direction-of-departure (DOD), direction-of-arrival (DOA), and polarization parameter estimation has been widely investigated for multiple-input multiple-output (MIMO) radars [1–3]. Apart from the aforementioned, many researchers have addressed the problem of direction estimation [4, 5]. For instance, a joint diagonalization-based method was proposed for estimating the 2D-DOD and 2D-DOA using the L-shaped array at both the transmitter and receiver in [6]. 2D-DOA estimates are obtained by adopting the estimation of signal parameters via rotational invariance techniques (ESPRIT) method to a MIMO system with a single receiving electromagnetic vector sensor (EVS) and multiple transmitting EVSs in [7]. For estimating 2D-DOA and polarization parameter, a mixed source localization method is devised with a linear tri-pole array in [8]. Besides, range-angle-dependent beam pattern is utilized to suppress range-ambiguous clutter and improve moving target detection performance in [9]. A joint range and angle estimation method is proposed for the MIMO radar with frequency diverse array (FDA-MIMO) by using the transmit sub-aperture optimization in [10]. The range and angle coupling are alleviated by forming the range and angle-decoupled beampattern.

In the paper, a method is proposed for estimating angle-polarization-range of targets using the EVSs in FDA-MIMO (EVS-FDA-MIMO) radar. This method is based on the spatial-invariant ESPRIT algorithm; it estimates angle-polarization without a priori knowledge of the sensor spacing.

T. Zhong (✉) · H. Tao
National Key Laboratory of Radar Signal Processing, Xidian University, Xi'an, China
e-mail: ttzhong@stu.xidian.edu.cn; hhtao@xidian.edu.cn

© The Author(s), under exclusive license to Springer Nature Switzerland AG 2024
S. Yeom (ed.), *4th International Conference on Electronics and Signal Processing*, Signals and Communication Technology,
https://doi.org/10.1007/978-3-031-36670-3_5

The paper is structured as follows: Section 2 presents the signal model for EVS-FDA-MIMO radar. In Sect. 3, the joint angle-polarization-range estimation problem is formulated. Besides, the ESPRIT-based method is used. The complexity analysis is addressed in Sect. 4. Extensive simulation results are presented in Sect. 5. Finally, conclusions and possible future research developments are discussed in Sect. 6.

2 Signal Model

Consider a collocated EVS-FDA-MIMO radar system consisting of N transmit EVSs and M receive EVSs placed according to a uniform linear array (ULA) configuration in both transmission and reception. The carrier frequency at the n - th element is

$$f_n = f_0 + (n-1)\,\Delta f, n = 1, 2, \cdots, N \tag{1}$$

where f_0 is the reference carrier frequency and a frequency increment Δf is introduced element by element in the transmit array with the first array EVS being the reference. The virtual data snapshot $\mathbf{y}_{s_k} \in \mathbb{C}^{36NM\times 1}$ of target signal can be expressed as

$$\mathbf{y}_{s_k} = \xi \mathbf{a}_{T_k} \circ \mathbf{a}_{R_k} \tag{2}$$

where $\circ$ denotes the Khatri-Rao product, ξ is the complex-valued reflection coefficient of the k - th target, and $\mathbf{a}_{T_k} \in \mathbb{C}^{6N\times 1}$ and $\mathbf{a}_{R_k} \in \mathbb{C}^{6M\times 1}$ are transmit and receive steering vectors, respectively, defined as

$$\begin{aligned} \mathbf{a}_{T_k} &= \mathbf{a}\left(r_{T_k}, \theta_{T_k}\right) \otimes \mathbf{c}_{T_k}\left(\theta_{T_k}, \varphi_{T_k}, \gamma_{T_k}, \eta_{T_k}\right) \\ \mathbf{a}_{R_k} &= \mathbf{b}\left(\theta_{R_k}\right) \otimes \mathbf{c}_{R_k}\left(\theta_{R_k}, \varphi_{R_k}, \gamma_{R_k}, \eta_{R_k}\right) \end{aligned} \tag{3}$$

where $\otimes$ denotes the Kronecker product and θ_{T_k} and θ_{R_k} stand for the elevation angle (measured from the normal direction to the target direction) in transmit and receive arrays, respectively. φ_{T_k} and φ_{R_k}, γ_{T_k} and γ_{R_k}, and η_{T_k} and η_{R_k} stand for azimuth angle, polarization auxiliary angle, and polarization phase difference, respectively. $\mathbf{a}\left(r_{T_k}, \theta_{T_k}\right) \in \mathbb{C}^{N\times 1}$ and $\mathbf{b}\left(\theta_{R_k}\right) \in \mathbb{C}^{M\times 1}$ are the steering vectors of spatial domain, respectively, expressed as

$$\mathbf{a}\left(r_{T_k}, \theta_{T_k}\right) = \mathbf{r}\left(r_{T_k}\right) \odot \mathbf{d}\left(\theta_{T_k}\right) \tag{4}$$

where the matrix $\mathbf{r}\left(r_{T_k}\right)$, $\mathbf{d}\left(\theta_{T_k}\right)$, and $\mathbf{b}\left(\theta_{R_k}\right)$ can be written as

$$\mathbf{r}\left(r_{T_k}\right) = \left[1, e^{-j2\pi\frac{2\Delta f}{c}r_{T_k}}, \cdots, e^{-j2\pi\frac{2\Delta f}{c}(N-1)r_{T_k}}\right]^{\mathrm{T}}$$
$$\mathbf{d}\left(\theta_{T_k}\right) = \left[1, e^{j2\pi\frac{d_T}{\lambda_0}\sin\left(\theta_{T_k}\right)}, \cdots, e^{j2\pi\frac{d_T}{\lambda_0}(N-1)\sin\left(\theta_{T_k}\right)}\right]^{\mathrm{T}} \quad (5)$$
$$\mathbf{b}\left(\theta_{R_k}\right) = \left[1, e^{j2\pi\frac{d_R}{\lambda_0}\sin\left(\theta_{R_k}\right)}, \cdots, e^{j2\pi\frac{d_R}{\lambda_0}(M-1)\sin\left(\theta_{R_k}\right)}\right]^{\mathrm{T}}$$

where $\odot$ is the Hadamard (element-wide) product and $(\cdot)^{\mathrm{T}}$ denotes the transpose. d_T and d_R are the spacing of transmit and receive elements, respectively. $\mathbf{d}\left(\theta_{T_k}\right)$ is the transmitting spatial steering vector, and $\mathbf{r}\left(r_{T_k}\right)$ is the transmitting range steering vector. As shown in (4)–(5), the transmit spatial steering vector is not only angle-dependent but also range-dependent. Therefore, the EVS-FDA-MIMO radar could utilize the transmit and receive degree of freedom (DOF) to determine the range and angle. The $\mathbf{c}(\theta, \varphi, \gamma, \eta)$ denotes an EVS spatial-polarization domain steering vectors of the transmitter and receiver. An EVS consists of three orthogonally oriented electric dipoles and three orthogonally oriented magnetic loops. The $\mathbf{c}(\theta, \varphi, \gamma, \eta)$ can be expressed as

$$\mathbf{c}(\theta, \varphi, \gamma, \eta) = [\mathbf{e}, \mathbf{h}]^{\mathrm{T}} = \mathbf{F}(\theta, \varphi)\,\mathbf{g}(\gamma, \eta) \quad (6)$$

where $\mathbf{e} = [e_x, e_y, e_z]$ is the electric field, $\mathbf{h} = [h_x, h_y, h_z]$ is the magnetic field, $\mathbf{F}(\theta, \varphi)$ represents 6×2 spatial angular location matrix in which $\theta \in [0, \pi)$, and$\varphi \in [0, 2\pi)$ and $\mathbf{g}(\gamma, \eta)$ represent 2×1 vector of polarization states in which $\gamma \in [0, \pi/2)$ and $\eta \in [-\pi, \pi)$. The matrix $\mathbf{F}(\theta, \varphi)$ and vector $\mathbf{g}(\gamma, \eta)$ can be written as

$$\mathbf{F}(\theta, \varphi) = \begin{bmatrix} \cos\theta\cos\varphi & -\sin\varphi \\ \cos\theta\sin\varphi & \cos\varphi \\ -\sin\theta & 0 \\ -\sin\varphi & -\cos\theta\cos\varphi \\ \cos\varphi & -\cos\theta\sin\varphi \\ 0 & \sin\theta \end{bmatrix} \quad \mathbf{g}(\gamma, \eta) = \begin{bmatrix} \sin\gamma e^{j\eta} \\ \cos\gamma \end{bmatrix} \quad (7)$$

The covariance matrix of the signal can be expressed as

$$\mathbf{R}_y = E\left\{\mathbf{y}^{(l)}\mathbf{y}^{(l)\mathrm{H}}\right\} = \mathbf{A}\mathbf{R}_\xi\mathbf{A}^H + \mathbf{R}_n \quad (8)$$

where $\mathbf{A} = \mathbf{A}_T \circ \mathbf{A}_R$ (i.e., $\mathbf{A}_R = \left[\mathbf{a}_{R_1}, \cdots, \mathbf{a}_{R_K}\right]$, $\mathbf{A}_T = \left[\mathbf{a}_{T_1}, \cdots, \mathbf{a}_{T_K}\right]$) denotes the joint angle-polarization-range steering vector of the EVS-FDA-MIMO radar, $\mathbf{R}_\xi = E\{\boldsymbol{\xi}^{(l)}\boldsymbol{\xi}^{(l)\mathrm{H}}\}$ is the target reflection coefficient covariance matrix, and $\mathbf{R}_n$ is the noise covariance matrix. Here, the covariance matrix $\mathbf{R}_y$ can be estimated from the finite number of snapshots L defined by

$$\hat{\mathbf{R}}_y = \frac{1}{L}\sum_{l=1}^{L}\mathbf{y}^{(l)}\mathbf{y}^{(l)\mathrm{H}} \quad (9)$$

3 Joint Angle-Polarization-Range Estimation

This section investigates parameter estimation in the EVS-FDA-MIMO radar and presents an approach for joint angle-polarization-range of transmitter and receiver estimation by exploiting the property of the EVS-FDA-MIMO radar, that is, according to the rotational invariance of the transmitting and receiving array.

3.1 *Estimation of the Unambiguous and Low Accurate Direction Cosines*

In this subsection, the rotational invariance of the receiving array is considered. The first and last $M - 1$ elements of the receiving array satisfy the spatial rotational invariance. From the matrix $\mathbf{A}$, the two data subsets associated with the receiving array can be obtained as

$$\begin{aligned} \mathbf{J}_2\mathbf{J}_1\mathbf{A} &= \mathbf{A}_{R_1}\mathbf{\Lambda}_R\mathbf{\Lambda}_{cT}\,(i,n) \\ \mathbf{J}_3\mathbf{J}_1\mathbf{A} &= \mathbf{A}_{R_2}\mathbf{\Lambda}_{cT}\,(i,n) \end{aligned} \tag{10}$$

where $i = 1, \cdots, 6, n = 0, \cdots, N - 1$. $\mathbf{A}_{R_1}$ and $\mathbf{A}_{R_2}$ denote the first and last $6(M - 1)$ rows of $\mathbf{A}_R$, respectively, and $\mathbf{\Lambda}_R = \text{diag}\,(\beta_1, \cdots, \beta_K)$ is receiving array phase shift matrix, $\mathbf{\Lambda}_{cT}\,(i,n) = \text{diag}\left(\alpha_1^n \mathbf{c}_{T_1}(i), \cdots, \alpha_K^n \mathbf{c}_{T_K}(i)\right)$ in which $\mathbf{c}_{T_k}(i)$ denotes the i - th row of $\mathbf{c}_{T_k}$. Then, define the following three selection matrices $\mathbf{J}_1$, $\mathbf{J}_2$, $\mathbf{J}_3$. According to the normalized Poynting vector, the unambiguous but low accurate estimates of the direction cosines can be obtained as

$$\begin{bmatrix} \hat{\mathbf{u}}_{R_k}^{\text{coarse}} \\ \hat{\mathbf{v}}_{R_k}^{\text{coarse}} \\ \hat{\mathbf{w}}_{R_k}^{\text{coarse}} \end{bmatrix} \triangleq \frac{\hat{\mathbf{e}}_{R_k}}{\left\|\hat{\mathbf{e}}_{R_k}\right\|} \times \frac{\hat{\mathbf{h}}_{R_k}^*}{\left\|\hat{\mathbf{h}}_{R_k}\right\|} = \begin{bmatrix} \sin\hat{\theta}_{R_k}\cos\hat{\varphi}_{R_k} \\ \sin\hat{\theta}_{R_k}\sin\hat{\varphi}_{R_k} \\ \cos\hat{\theta}_{R_k} \end{bmatrix} \tag{11}$$

where $\hat{\mathbf{e}}_{R_k}$ and $\hat{\mathbf{h}}_{R_k}$ can be obtained from the estimate eigenvalues of $\hat{\mathbf{c}}_{R_k} \in C^{6\times 1}$ and $\times$ denotes the vector cross product.

3.2 *Estimation of the Unambiguous and High Accurate Angle Parameters*

Considering estimation of the unambiguous and high accurate elevation, the receiving array divides into the first $M - 1$ EVSs and last $M - 1$ EVSs. For the k - th target signal, the rotational invariance of the receiving array can be expressed as

$$\mathbf{J}_{R_2}\mathbf{b}\left(\theta_{R_k}\right) = q_{R_k}\mathbf{J}_{R_1}\mathbf{b}\left(\theta_{R_k}\right) \tag{12}$$

where $q_{R_k} = e^{j2\pi \frac{d_R}{\lambda_0}\sin(\theta_{R_k})}$, $\mathbf{J}_{R_1} = \left[\mathbf{I}_{M-1}\mathbf{0}_{(M-1)\times 1}\right]$, and $\mathbf{J}_{R_2} = \left[\mathbf{0}_{(M-1)\times 1}\mathbf{I}_{M-1}\right]$ denote the selection matrices. The (12) is extended to the joint angle-polarization-range steering vector

$$\mathbf{a}_{T_k} \otimes \left[\mathbf{J}_{R_2}\mathbf{b}\left(\theta_{R_k}\right)\right] = q_{R_k}\mathbf{a}_{T_k} \otimes \left[\mathbf{J}_{R_1}\mathbf{b}\left(\theta_{R_k}\right)\right] \tag{13}$$

Then, the high accurate but cyclically ambiguous z-axis direction cosine estimates as

$$\hat{\mathbf{w}}_{R_k}^{\text{fine}} = -\frac{\lambda_0 \arg\left\{\left[\hat{\mathbf{\Phi}}_{Rf}\right]_k\right\}}{2\pi d_R} \tag{14}$$

where $[\cdot]_k$ denotes the (k, k) - th element of a matrix and $k = 1, \cdots, K$ and $\arg(\cdot)$ denote the phase angle of a complex number in the interval $[-\pi, \pi]$. $\hat{\mathbf{\Phi}}_{Rf}$ is estimates of $\mathbf{\Phi}_{Rf} = \text{diag}\left[q_{T_1}, q_{T_2}, \cdots, q_{T_K}\right]$. Therefore, by using (11), (14) and the disambiguation approach can obtain final estimation of the unambiguous and high accurate z-axis direction cosine as

$$\hat{\mathbf{w}}_{R_k}^{\text{final}} = \hat{\mathbf{w}}_{R_k}^{\text{fine}} + \mathring{m}_{R_k}\lambda/\text{d}_R \tag{15}$$

where $\mathring{m}_{R_k} = \underset{m_{R_k}}{\arg\min}\left|\hat{\mathbf{w}}_{R_k}^{\text{coarse}} - \hat{\mathbf{w}}_{R_k}^{\text{fine}} - m_{R_k}\lambda/\text{d}_R\right|$ is the cyclically ambiguous reference value, for $m_{R_k} \in \left\{\left\lceil d_R/\lambda\left(-1 - \hat{\mathbf{w}}_{R_k}^{\text{coarse}}\right)\right\rceil, \left\lfloor d_R/\lambda\left(1 - \hat{\mathbf{w}}_{R_k}^{\text{coarse}}\right)\right\rfloor\right\}$, and $\lceil\cdot\rceil$ and $\lfloor\cdot\rfloor$ denote the ceil function and floor function, respectively. Finally, the accurate elevation angle can be derived as

$$\hat{\theta}_{R_k}^{\text{final}} = \arcsin\left(\hat{\mathbf{w}}_{R_k}^{\text{final}}\right) \tag{16}$$

3.3 Estimation of Polarization Parameters

Substituting the estimated parameter elevation and $\hat{\mathbf{c}}_{R_k}$ into (6), the polarization parameters can be estimated as

$$\hat{\gamma}_{R_k} = \arctan\left(\frac{\left|\hat{\mathbf{g}}_1\left(\gamma_{R_k}, \eta_{R_k}\right)\right|}{\left|\hat{\mathbf{g}}_2\left(\gamma_{R_k}, \eta_{R_k}\right)\right|}\right) \quad \hat{\eta}_{R_k} = \angle\left(\frac{\hat{\mathbf{g}}_1\left(\gamma_{R_k}, \eta_{R_k}\right)}{\hat{\mathbf{g}}_2\left(\gamma_{R_k}, \eta_{R_k}\right)}\right) \tag{17}$$

where $\hat{\mathbf{g}}\left(\gamma_{R_k},\eta_{R_k}\right)\triangleq\left[\hat{\mathbf{g}}_1\left(\gamma_{R_k},\eta_{R_k}\right),\hat{\mathbf{g}}_2\left(\gamma_{R_k},\eta_{R_k}\right)\right]^{\mathrm{T}}=\left(\hat{\mathbf{F}}^{\mathrm{H}}\left(\theta_{R_k},\varphi_{R_k}\right)\hat{\mathbf{F}}\left(\theta_{R_k},\varphi_{R_k}\right)\right)^{-1}$ $\hat{\mathbf{F}}^{\mathrm{H}}\left(\theta_{R_k},\varphi_{R_k}\right)\hat{\mathbf{c}}_{R_k}$.

3.4 *Estimation of Range*

In this subsection, consider the rotational invariance of the transmitting array. The first and last $N-1$ EVSs of the transmitting array satisfy spatial rotational invariance. For the k - th target signal, the rotational invariance of the transmitting array can be expressed as

$$\mathbf{J}_{T_2}\mathbf{a}\left(r_{T_k},\theta_{T_k}\right)=h_{T_k}\mathbf{J}_{T_1}\mathbf{a}\left(r_{T_k},\theta_{T_k}\right) \tag{18}$$

where $h_{T_k}=e^{j2\pi\frac{d_T}{\lambda_0}\sin\left(\theta_{T_k}\right)}e^{-j2\pi\frac{2\Delta f}{c}r_{T_k}}$, $\mathbf{J}_{T_1}=\left[\mathbf{I}_{N-1}\mathbf{0}_{(N-1)\times 1}\right]$, and $\mathbf{J}_{T_2}=\left[\mathbf{0}_{(N-1)\times 1}\mathbf{I}_{N-1}\right]$ are selection matrices. Note that the transmitting steering vector $\mathbf{a}\left(r_{T_k},\theta_{T_k}\right)$ contains not only elevation angle but also range parameter. It follows that elevation angle and range parameter cannot be obtained directly using the rotational invariance of the transmitting array. Assuming $\theta_{T_k}=\theta_{R_k}$, since elevation angle accurate estimation has been obtained for the receiving array, range parameter can be estimated when the elevation angle accurate estimation is substituted into (18). It is extended to the joint transmit-receive steering vector

$$\mathbf{a}_{R_k}\otimes\left[\mathbf{J}_{T_2}\mathbf{a}\left(r_{T_k},\theta_{T_k}\right)\right]=h_{T_k}\mathbf{a}_{R_k}\otimes\left[\mathbf{J}_{T_1}\mathbf{a}\left(r_{T_k},\theta_{T_k}\right)\right] \tag{19}$$

After the accurate estimated angle value is substituted into $\mathbf{\Phi}_{TR}$, range estimation can be calculated as follows:

$$\hat{r}_{T_k}=\frac{\left\{2\pi d_T\sin\left(\hat{\theta}_{T_k}^{\text{final}}\right)/\lambda_0-\arg\left\{\left[\hat{\mathbf{\Phi}}_{TR}\right]_k\right\}\right\}c}{4\pi\Delta f} \tag{20}$$

where $\hat{\mathbf{\Phi}}_{TR}$ is the estimates $\mathbf{\Phi}_{TR}=\operatorname{diag}\left[h_{T_1},h_{T_2},\cdots,h_{T_K}\right]$.

3.5 *Pair Matching of Spatial-Polarization Angle and Range*

It should be noted that the diagonal elements in the same position of $\mathbf{\Phi}_{Rf}$ and $\mathbf{\Phi}_{TR}$ may not correspond to the same target, but there is no mismatch between the diagonal element and its corresponding eigenvector. To pair the angle and range successfully, the method in [11] can be employed.

Table 1 Analysis of the computational complexity for two methods

Method	Computational complexity
EVS-MIMO	$\mathcal{O}\left((36NM)^2L + (36NM)^3 + 12K + (12(M-1)(N-1) + 7(M+N) + (36MN-K)(36MN+1))K^2 + 6K^3\right)$
Proposed	$\mathcal{O}\left((36NM)^2L + (36NM)^3 + 6K + (72NM - 36N - 17M - 12)K^2 + 36K^3\right)$

Table 2 Simulation parameters of the EVS-FDA-MIMO radar

Parameters	Values
Transmit elements	$N = 10$
Receive elements	$M = 10$
Target elevation angle	$\theta_{T_k} = \theta_{R_k} = 30°$
Sensor spacing	$d_T = d_R = \lambda_0/2$
Waveform bandwidth	$B = 15$MHz
Pulse repetition frequency	$f_{\mathrm{PRF}} = 5000$Hz
Reference frequency	$f_0 = 10$GHz
Frequency increment	$\Delta f = 0.3$MHz

4 Complexity Analysis

In this section, the assessment of computational burden involved by the proposed estimators and the EVS-MIMO radar [12] are provided. Table 1 shows the computational complexity of two methods, where N and M signify the number of transmitting and receiving array EVSs, respectively. L denotes the number of snapshots and K symbolizes the number of targets.

5 Simulation Results

In this section, two examples are provided to access the proposed method performance and to estimate target angle, polarization, and range with reference to an EVS-FDA-MIMO radar system. The values of the simulation parameters involved in the analyzed case studies are listed in Table 2, and target parameters are $\{\theta = 30°, \varphi = 70°, \gamma = 40°, \eta = 80°, r = 40\text{km}\}$. For comparison, both the EVS-MIMO radar and the EVS-FDA-MIMO radar use a ULA in transmitter and receiver. The root mean square error (RMSE) is used as the performance metric. The RMSE of θ is defined as

$$\mathrm{RMSE}_\theta = \sqrt{\frac{1}{KM_0}\sum_{k=1}^{K}\sum_{m_0=1}^{M_0}\left\|\hat{\theta}_k^{m_0} - \theta_k\right\|^2} \tag{21}$$

where M_0 denotes the number of Monte Carlo experiments and $\hat{\theta}_k^{m_0}$ denotes the estimate of θ_k in the m_0 - th Monte Carlo experiment. $\text{RMSE}_{\theta,\,\text{coarse}}$, $\text{RMSE}_{\theta,\,\text{fine}}$, RMSE_r, RMSE_γ, and RMSE_η are defined similarly.

5.1 RMSE Analysis with Respect to Different Input SNR

In the first experiment, consider influence of signal-to-noise ratio (SNR) in the EVS-MIMO radar, the EVS-FDA-MIMO radar, and the Cramér-Rao bound (CRB). The result of RMSE versus SNR is drawn in Fig. 1. To this end, 200 Monte Carlo simulations have been carried out. During the Monte Carlo trials, the number of snapshots is 200, SNR is −20~20 dB with 5 dB intervals, and noise is AGWN.

Note that CRB for angle-polarization in the EVS-MIMO radar is lower than that for the EVS-FDA-MIMO radar due to the increased aperture of the virtual

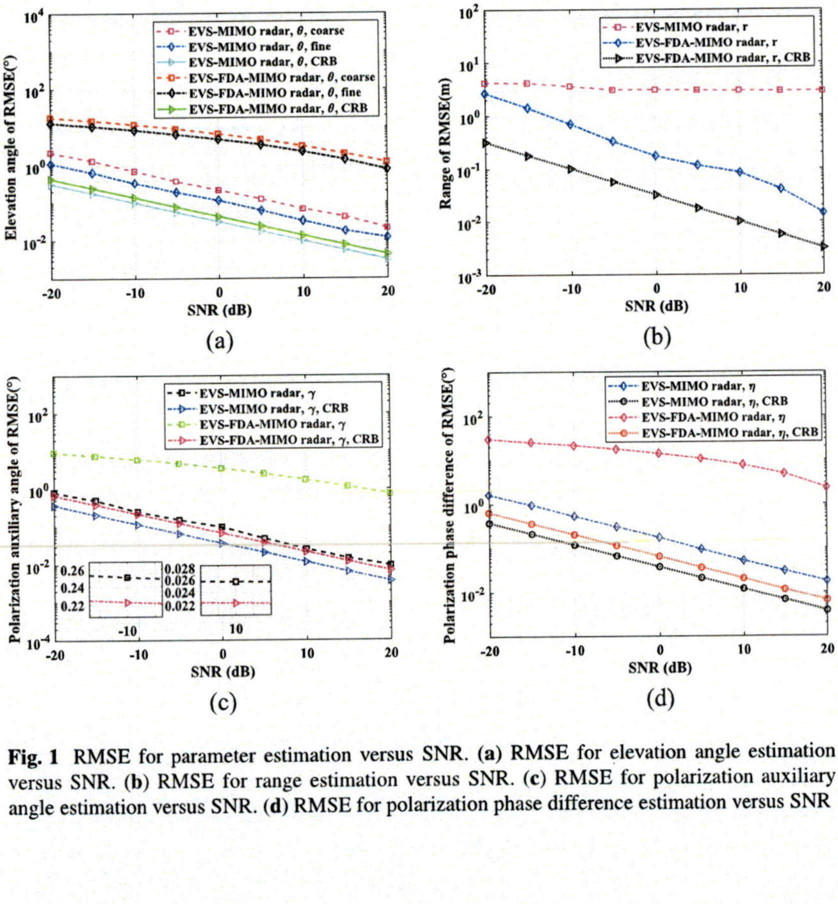

Fig. 1 RMSE for parameter estimation versus SNR. (**a**) RMSE for elevation angle estimation versus SNR. (**b**) RMSE for range estimation versus SNR. (**c**) RMSE for polarization auxiliary angle estimation versus SNR. (**d**) RMSE for polarization phase difference estimation versus SNR

array in the EVS-MIMO radar. Furthermore, the main reason for the unsatisfactory performance of the EVS-FDA-MIMO radar at low SNRs is that it only uses the information inside of the EVS but does not make use of the array aperture for elevation angle coarse estimation, which leads to poor performance of coarse parameter estimation as can be observed in Fig. 1, and the incorrect coarse estimation at low SNRs will make the disambiguation process fail. For elevation angle fine estimation, since the elevation angle and range of the target are coupled in transmit dimension, performance loss of fine parameter estimation is caused. It is seen from Fig. 1b that the range estimation accuracy can be enhanced with the increase of SNR in the EVS-FDA-MIMO radar. However, for the EVS-MIMO radar, the accuracy cannot be improved since the range resolution is determined by the bandwidth of the transmitted waveform, which is fixed in the EVS-MIMO radar.

5.2 RMSE Analysis with Respect to Different Input Snapshots

In the second experiment, consider influence of snapshots in the EVS-MIMO radar, the EVS-FDA-MIMO radar, and the CRB. The result of RMSE versus snapshots is drawn in Fig. 2. To this end, 200 Monte Carlo simulations have been carried out. During the Monte Carlo trials, the number of snapshots is 200~2000 with 200 intervals, SNR is 20 dB, and noise is AGWN.

As shown in Fig. 2, the EVS-FDA-MIMO radar is able to provide the DOF in range domain, the range estimation accuracy is improved evidently, and the range RMSE of the EVS-FDA-MIMO radar is very close to CRB. Moreover, the RMSE of range estimates are much smaller in the EVS-FDA-MIMO radar than those in the EVS-MIMO radar. Note that its performance is not as good as the EVS-MIMO radar at lower SNRs; the reason is the same as that of the first experiment. Therefore, the performance loss of the EVS-FDA-MIMO radar is suffered in angle-polarization estimation, and the super-resolution in range is provided.

6 Conclusion

In this paper, the problem of target angle-polarization-range estimation with the EVS-FDA-MIMO radar has been investigated. At the design stage, an estimator, with ESPRIT-based method using the rotational invariance, has been devised. First, the unambiguous and low accurate direction cosines are estimated. Second, the unambiguous and high accurate elevation angle is obtained. Due to the constraint of uniform linear array, only the high accurate and unambiguous estimation of one-dimensional spatial angle can be estimated. Then, estimation of polarization parameters and range are devised. Furthermore, the pair matching of angle-polarization-range and the complexity computation analysis are considered. Finally, simulation results validate the performance of the proposed EVS-FDA-MIMO

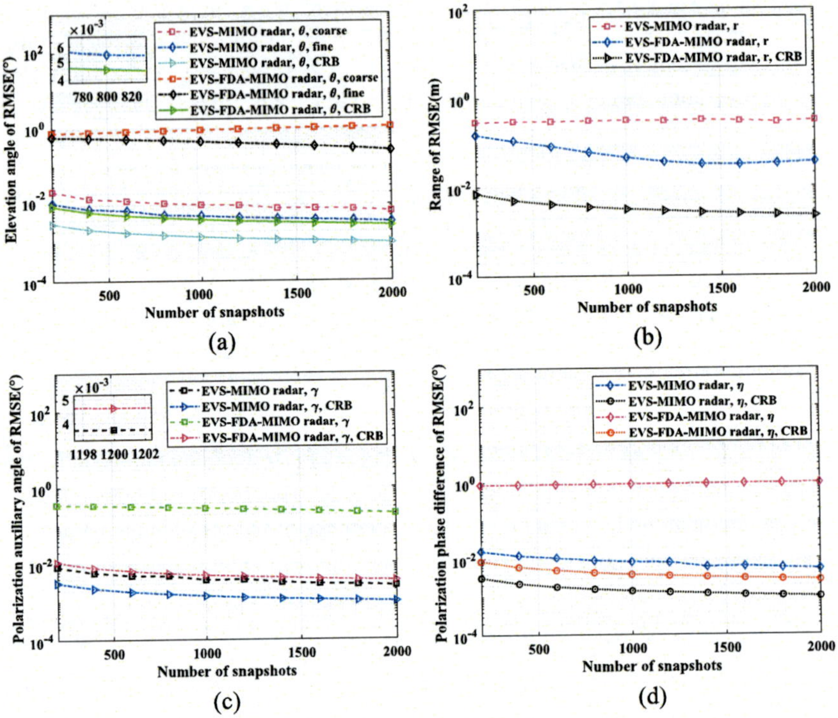

Fig. 2 RMSE for parameter estimation versus snapshots. (**a**) RMSE for elevation angle estimation versus snapshots. (**b**) RMSE for range estimation versus snapshots. (**c**) RMSE for polarization auxiliary angle estimation versus snapshots. (**d**) RMSE for polarization phase difference estimation versus snapshots

radar technique. Compared with some existing methods, the proposed method can measure both the targets' spatial and polarization information, extends the array aperture without additional antennas, and, therefore, is able to achieve high-accuracy range estimation performance. Additionally, it avoids spectral search. It is worth mentioning that the EVS-FDA-MIMO radar can also be extended to the scenario that arrays are in linear or nonlinear frequency increments.

References

1. Xie, R., Liu, Z., & Wu, J. (2012). Direction finding with automatic pairing for bistatic MIMO radar. *Signal Processing, 92*(1), 198–203.
2. Zheng, G., & Tang, J. (2017). DOD and DOA estimation in bistatic MIMO radar for nested and coprime array with closed-form DOF. *International Journal of Electronics, 104*(5), 885–897.

3. Zheng, G., & Chen, B. (2015). Unitary dual-resolution ESPRIT for joint DOD and DOA estimation in bistatic MIMO radar. *Multidimensional Systems and Signal Processing, 26*(1), 159–178.
4. Wong, K. T., & Zoltowski, M. D. (1997). Uni-vector-sensor ESPRIT for multisource azimuth, elevation, and polarization estimation. *IEEE Transactions on Antennas Propagation, 45*(10), 1467–1474.
5. Lan, L., Rosamilia, M., Aubry, A., De Maio, A., & Liao, G. (2021). Single-snapshot angle and incremental range estimation for FDA-MIMO radar. *IEEE Transactions on Aerospace and Electronic Systems, 57*(6), 3705–3718.
6. Xia, T. (2015). Joint diagonalization based 2D-DOD and 2D-DOA estimation for bistatic MIMO radar. *Signal Processing, 116*(7–12), 7.
7. Gu, C., He, J., Li, H., & Zhu, X. (2013). Target localization using MIMO electromagnetic vector array systems. *Signal Processing, 93*(7), 2103–2107.
8. Ma, H., Tao, H., & Xie, J. (2020). Mixed far-field and near-field source localization using a linear tripole array. *IEEE wireless Communications Letters, 9*(6), 889–892.
9. Baizert, P., Hale, T. B., Temple, M. A., & Wicks, M. C. (2006). Forward-looking radar GMTI benefits using a linear frequency diverse array. *Electronics Letters, 42*(22), 1311–1312.
10. Wang, W., & So, H. C. (2014). Transmit subaperturing for range and angle estimation in frequency diverse array radar. *IEEE Transactions Signal Processing, 62*(8), 2000–2011.
11. Wong, K. T., & Zoltowski, M. D. (2000). Self-initiating MUSIC-based direction finding and polarization estimation in spatio-polarizational beamspace. *IEEE Transactions on Antennas and Propagation, 48*(8), 1235–1245.
12. Chintagunta, S., & Ponnusamy, P. (2018). 2D-DOD and 2D-DOA estimation using the electromagnetic vector sensors. *Signal Processing, 147*(163–172), 163.

An Enhanced Approach for Environmental Sound Classification Using Multi-window Augmentation

Krishna Presannakumar and Anuj Mohamed

1 Introduction

Environmental sound classification (ESC) research, which focuses mostly on detecting particular sound occurrences, such as identifying dog barking, gunshots, and air conditioning noises, has attracted more and more interest. Numerous real-world applications, such as soundscape monitoring, home automation, audio surveillance systems, and robotic hearing, have made use of the research related to ESC. The environmental sound differs from regular and structured sounds like speech and music because it lacks both phonetics and static temporal variations like melodies and rhythms. As a result, it is challenging to identify common features that can accurately reflect a wide range of temporal patterns. Additionally, the environmental sound comprises a lot of noise and some sounds unrelated to the sound event, which results in a complex composition structure with variety, complexity, and attributes that are not structured. Recently deep learning models are becoming popular in the research of sounds. Deep learning models perform efficiently and accurately if the dataset is rich and adequate. By creating additional and distinct variants for training datasets, data augmentation helps deep learning models to perform better and produce good results. The convolutional neural network is the most popularly used deep learning model and it can learn in both the time domain and frequency domain at the same time [1].

In 2021, Lam Pham et al. [1] suggested an encoder-decoder network model for acoustic scene classification (ASC), which employed several low-level spectrogram features at the front-end, which are then turned into higher-level features by a trained CNN-DNN front-end encoder. In 2020, Jie Xie [2] proposed to divide the audio

K. Presannakumar (✉) · A. Mohamed
School of Computer Sciences, Mahatma Gandhi University, Kottayam, Kerala, India
e-mail: anujmohamed@mgu.ac.in

© The Author(s), under exclusive license to Springer Nature Switzerland AG 2024

S. Yeom (ed.), *4th International Conference on Electronics and Signal Processing*, Signals and Communication Technology,
https://doi.org/10.1007/978-3-031-36670-3_6

signal into frames using a sliding window and to learn a representation from the raw audio signal using a 1D CNN-LSTM network. In 2018, Shaobo Li et al. [3] proposed a unique stacked CNN model with numerous convolutional layers with decreasing filter sizes to enhance the performance of CNN models with either Log-Mel feature input or raw waveform input. In 2021, Tao Zhang et al. [4] introduced a method utilizing Gradient-weighted Class Activation Mapping (Grad-CAM), a CNN visualization approach that analyzes what kind of information is received by a CNN.

Sometimes it is challenging to meet the demands of real-time performance and deploy CNN to embedded devices with limited storage and processing power, because of the slow running speed caused by the numerous parameters and calculations [5]. Therefore, to lower the operating cost of CNN and increase the classification speed of environmental sound, we propose a gated CNN, which is a combination of long short-term memory (LSTM) and CNN. There are many data augmentation techniques that are commonly used in image processing such as padding, zooming, cropping, rescaling, etc. Even though these can be applied to audio as well but most of the time when image augmentation is applied to audio, it will significantly alter the spectrogram and the sound that it represents. To address the unavailability of data and overfitting-related problems, the research demonstrates an ESC system based on an enhanced multi-window spectrogram augmentation approach and a gated convolutional network. First, the approach makes use of the idea of multiple windows in terms of both the size (overlapping values) and the type and uses this on audio signals to constantly find audio spectrograms needed for training. Then, using the generated training datasets, networks are adjusted for label recognition and fine-tuned to create a model resilient to audio degradation brought on by noise. Our approach has three benefits: By using a multi-window approach, we can get features with smaller and more extended sounds, which are crucial for environmental sounds. It solves the unavailability of data to develop a complicated through producing additional data from the learning model. In the audio feature extraction phase, using a multi-window strategy for finding the ideal window size is a crucial problem. Since the approach used a gated convolutional network, which is the combination of CNN and long short-term memory networks (LSTMs), it resolves the problem of overfitting and vanishing gradient.

2 Proposed Methodology

This section explains the proposed environment sound classification dataset, the methods used for multi-window data augmentations, and the proposed architecture of gated convolutional neural network and performance evaluation methods. The urbansound8k and ESC dataset is used as the primary data and it goes through preprocessing steps, such as unifying audio channels, sampling rate, and size before using the audio clips to generate spectrograms. A multi-window data augmentation

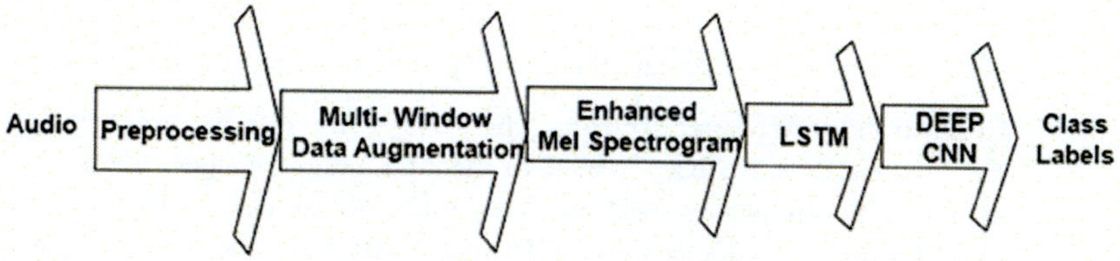

Fig. 1 Proposed architecture for environmental sound classification

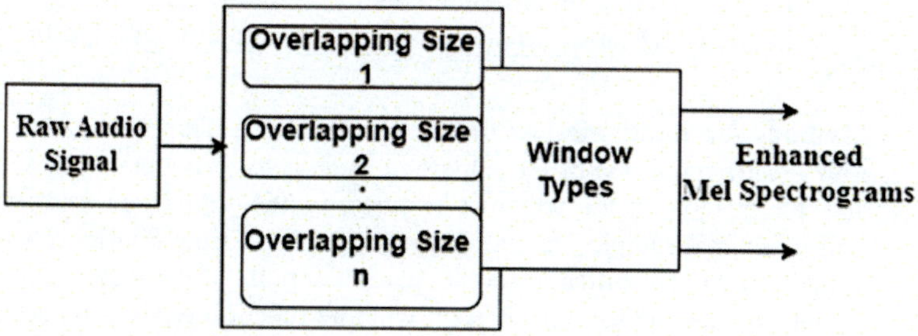

Fig. 2 Proposed method for enhanced multi-window spectrogram augmentations

and a gated convolutional neural network (LSTM-CNN) model are the two phases of the suggested methodology. By using various window sizes, overlapping values, and types to extract audio features, the first stage produces more audio data for training the model. The retrieved features are used for audio classification in the gated CNN model's second stage. Figure 1 shows the proposed model architecture.

2.1 Multi-window Data Augmentations

In audio classification using spectrograms, the features are retrieved using a windowed signal, often known as a frame. An audio signal often only remains steady for a brief period, and the size of the windowing function directly affects the feature values that can be retrieved from the audio signal. For instance, if the chosen window length is too short, we might not have enough data samples to obtain a trustworthy spectral analysis of the signal; conversely, if it is too long, the signal varies excessively across the frame. In light of this, choosing the width of a windowing function is an important step and is challenging when prior knowledge about the input signal is unknown [5]. Figure 2 illustrates the proposed method for enhanced multi-window spectrogram augmentations

According to experiments, choosing the best window size improves the association between a speech signal's acoustic representation and how it is perceived by listeners. Window width (in milliseconds), window offset or overlap (in milliseconds), and window shape are the three tuples that make up a window function.

Generation of Mel spectrogram The Mel spectrogram is a spectrogram in which the frequencies have been scaled to the Mel range. A spectrogram is a graphic depiction of a frequency spectrum. They reflect the signal's volume or strength with time at different frequencies. Similar to a spectrogram, the Mel spectrogram is also described as a depiction of the short-term power spectrum of a sound stream. The main distinction is that the Mel spectrogram's y-axis uses the Mel scale rather than a spectrogram's log scale of frequencies [4, 5]. Different types of windows can be selected and using these windows we can generate different spectrograms for the audio signal.

Methods for multi-window data augmentations During the generation of the Mel spectrogram, one of the important steps is to select the window type and the size for generating the windows (frames) of the signals. In each window, there will be several samples, and after applying a short time Fourier transforms to this window, it will be shifted to another particular point of the signal. Depending on the signal, we can use several different window function types. An authentic window plot illustrates that a window's frequency characteristic is a continuous spectrum with a main lobe and multiple side lobes. The side lobes approach zero, and the main lobe is centered at each frequency component of the time-domain signal. The size of the side lobes shows how the windowing function affects frequencies close to the main lobes. A nearby weak sinusoidal signal's main lobe performance can be drowned out by a powerful sinusoidal signal's side lobe response. Lower side lobes typically increase the bandwidth of the principal lobe while decreasing leakage in the measured FFT. The pace at which the side lobe peaks exponential decay is known as the side lobe roll-off rate. Spectral loss can be decreased by raising the side lobe roll-off rate.

It is not easy to choose a window function. Each window function has unique properties that make it suitable for various applications. We must determine the frequency content of the signal to choose a window function. Since our research uses environmental audio that also contains background noises, the window length should be both short enough to ensure that the windowed signal is practically stationary across the window interval and long enough to ensure that a decent frequency resolution is provided by the windowed signal's Fourier transform. The number of windows (NW) in a signal is calculated using Eq. (1) according to the length of audio (A_1) and the sampling frequency (Sf).

$$Nw = \frac{\sum_{i=1}^{n} (A_1)}{s_f \times O_i} \tag{1}$$

where O_i is the overlapping parameter and n is the total number of overlapping parameters. In our research, we are using three values for overlapping parameters 25 ms, 50 ms, and 100 ms. Once the number of windows is calculated, the window size (number of samples) is selected using Eq. (2).

$$\omega_s = \frac{A_l}{O_i} \tag{2}$$

where Al length of audio O_i is the overlapping parameter. In our research we are using hamming, Kaiser Bessel-derived, hann window types along with window sizes 25 ms, 50 ms, and 100 ms. The different size of windows is used to extract features with an overlap of 15 ms window. Algorithm 1 explains the steps involved in the proposed model.

Algorithm 1 Enhanced Multi-window Spectrogram Augmentation (EMWS-Aug)

Input: Raw audio signal in waveform

Output: Enhanced Mel spectrograms

Let A = {*Audio clips in the dataset*}, *Al = length of audio, N = Number of audios in dataset*

1. Repeat the following steps for every audio in A.
2. Calculate the number of windows for the audio signal using Eq. (1).
3. Calculate window size for audio using Eq. (2) for each overlapping value.
4. For each overlapping value, select the window type and generate a Mel spectrogram for each type.

The total number of Mel spectrograms for each audio signal will be generated according to the type of *window* and the number of overlapping values.

$$\text{Number of Mel spectrogram} = N\text{w}_\text{T} \times O_\text{T} \tag{3}$$

where N_{WT} is the number of window types and O_T is the total number of overlapping values. According to the size of the dataset and the application, we can select the types of windows and overlapping values.

2.2 *Gated Convolutional Neural Network*

This research uses a combination of long short-term memory (LSTM) and convolutional neural network (CNN). There are different types of recurrent convolutional neural networks that can be used; we choose LSTM because of its ability to build long-term dependencies. Convolutional neural networks, also referred to as ConvNets, provide the foundation for the majority of deep learning tasks. There are input, output, and several hidden layers in CNNs, at least one of which is a 2D convolutional layer that applies a mathematical linear function. The main layer of a CNN is a convolutional layer, which does the majority of the computational load. Rectified Linear Unit (ReLU) layer does the process of converting weighted input into output via an activation function. The usage of a pooling layer in CNN helps to reduce the spatial size of the convolved feature, which lowers the amount of processing power needed to handle the data. Additionally, it lessens overfitting, in the CNN. All of the neurons in a fully connected layer (FC) are connected, as suggested by the term [5]. The long short-term memory (LSTM) network is a type of artificial recurrent neural network. There are typically feedforward connections in regular neural networks; on the other hand, LSTM networks have feedback connections. Gated cells in LSTMs will learn what to accept and what to reject by backpropagation of error, guessing iteratively, and gradient descent weight adjustment.

The LSTM receives the EMWS-augmented Mel spectrogram as the input and performs feature extraction. These features are passed into the CNN architecture for further classification. The LSTM uses a linear layer and dropout layer to avoid overfitting. The final layer of CNN contains the same number of neurons as the number of classes in the datasets. The model training is conducted for 50 epochs and the batch size is set to 32. At each of the training epochs, the training pairs are reshuffled. The performance is evaluated at each epoch. Accuracy is computed using Eq. (4).

$$\text{Accuracy} = \frac{T_P + T_N}{T_P + T_N + F_p + F_N} \times 100 \quad (4)$$

where T_P is true positive, T_N is true negative, F_P is false positive, and F_N is false negative. The CNN model uses a batch SGD for training the model and is much faster because it only requires a smaller number of computations. The number of computations is less when each iteration only uses a small batch of the dataset. This leads to better performance of the system as it avoids the local minima and prevents overfitting due to the small size of the data being used during the iteration.

3 Results and Discussions

The effectiveness of the developed systems was evaluated using the accuracy or classification rate per audio recording, which is the proportion of correctly classified files to the total number of tested files. In the experiments, the Urban8k and ESC datasets were used. The suggested enhanced multi-window gated CNN model effectively generates the augmented dataset and performs the classification with more efficiency. The overfitting problem and vanishing gradient problem are solved by using a combination of LSTM and CNN, which performs much better compared to other networks. Figure 3 shows the Mel spectrograms generated using different window types.

The Mel spectrogram, generated using different types of windows and overlapping values, together with a combined LSTM-convolutional neural network, was used in the tests on the Urban8k and ESC datasets. The mean accuracy of different approaches is illustrated in Table 1.

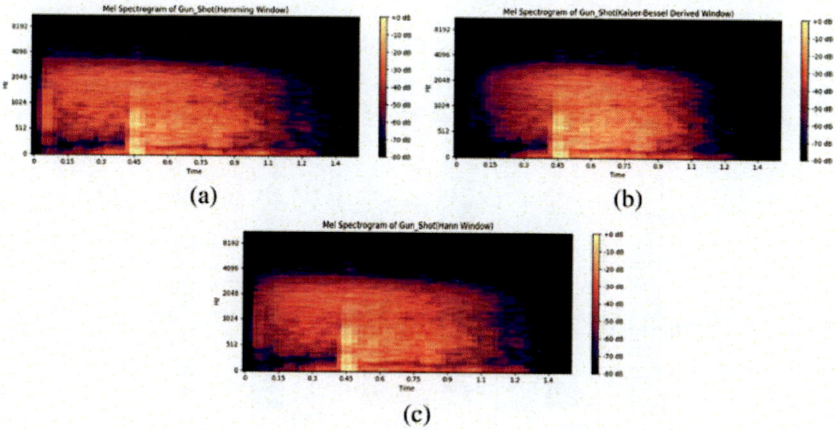

Fig. 3 Mel spectrogram generated by different window types. (**a**) Hamming. (**b**) Kaiser Bessel derived. (**c**) Hann

Table 1 Accuracy of different approaches

	UrbanSound8k	ESC-50	ESC-10
Shaobo Li et al. [3]	90.2%	81.1%	91.4%
Boddapati et al. [6]	93%	86%	73%
Salamon et al. [7]	79%	–	72%
Piczak et al. [8]	73%	60.5%	80.50%

Table 2 Evaluation metrics for proposed models

Model		Accuracy	Precision	Recall	F1 score
Urban8k	EMWS-aug + LSTM	0.9010	0.8755	0.9078	0.8457
	EMWS-aug + CNN	0.9248	0.8911	0.9254	0.8841
	EMWS-aug + LSTM+CNN	0.9381	0.9018	*0.9577*	0.9155
ESC 10	EMWS-aug + LSTM	0.9111	0.8475	0.8948	0.8564
	EMWS-aug + CNN	0.9287	0.9214	0.9379	0.9217
	EMWS-aug + LSTM+CNN	*0.9497*	*0.9517*	0.9474	*0.9301*
ESC 50	EMWS-aug + LSTM	0.8815	0.8997	0.8814	0.8990
	EMWS-aug + CNN	0.9019	0.9145	0.9248	0.9174
	EMWS-aug + LSTM+CNN	0.9175	0.9341	0.9422	0.9270

The dataset is divided into two parts: 20% is used for testing, while the remaining 80% is used for training. The evaluation of the model is performed using the entire testing and training dataset. The model's primary purpose is to identify environmental sounds with high accuracy. The experimental results predicted by different models are illustrated in Table 2 and Fig. 4.

The model uses categorical cross entropy as the loss function and gradient descent optimization. The learning rate, epochs, and batch size are adjusted as part of the hyperparameter tuning. The learning rate of the model is adjusted to 0.0001 to

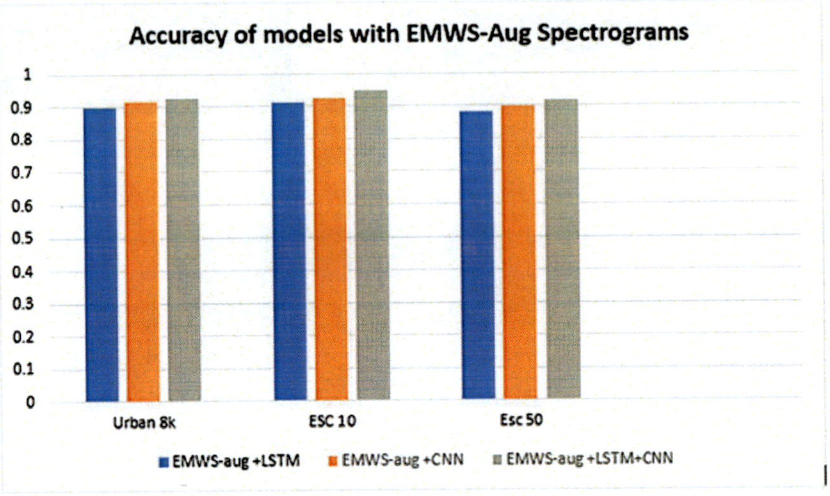

Fig. 4 Accuracy of EMWS-Aug gated CNN model for Urban8k dataset

increase accuracy. The models were subjected to experiments across 25, 50, and 70 epochs. There is a set batch size of 32. The training loss was significant at first, but as the number of epochs increased, the loss gradually decreased and stabilized after 40 epochs. After 15 epochs, the testing accuracy grew, and after 35 epochs, it stabilized. At the stage of epoch 50, the model has the highest accuracy; thus the model is trained up to 50 epochs. Compared to the other models, this model produces better outcomes. On test data from the Urban8k, ESC 10, and ESC 50 datasets, the model's overall accuracy is 93.81%, 94%, and 91.75% respectively. The evaluation shows that the model performs better on used datasets; on all benchmarks we convincingly beat the previous state of the art with more accuracy and efficiency.

4 Conclusion

The research mainly concentrated on how to solve the challenges relating to data availability and audio degradation brought on by background noises during the classification. The research attempts to address these concerns via data augmentation of the audio signal and has produced a promising result. The data augmentation is performed using the proposed enhanced multi-window augmentation method, which generated more Mel spectrograms using different types of windows and the number of overlapping values. The changes in overlapping values improved the quality of the spectrogram by extracting all sounds in the signals. The accuracy of the method is further enhanced by deploying a combination of LSTM and CNN for the classification of enhanced Mel spectrograms. Consequently, the solution removes the issues related to overfitting and the vanishing gradient problem. The proposed method is providing better results for a deep learning strategy to identify the source of the audio signals.

References

1. Pham, L., Phan, H., Nguyen, T., Palaniappan, R., Mertins, A., & McLoughlin, I. (2021). Robust acoustic scene classification using a multi-spectrogram encoder-decoder framework. *Digital Signal Processing, 110*, 102943.
2. Xie, J., Hu, K., Zhu, M., & Guo, Y. (2020). Bioacoustic signal classification in continuous recordings: Syllable-segmentation vs sliding-window. *Expert Systems with Applications, 152*, 113390.
3. Li, S., Yao, Y., Hu, J., Liu, G., Yao, X., & Hu, J. (2018). An ensemble stacked convolutional neural network model for environmental event sound recognition. *Applied Sciences, 8*, 1152.
4. Zhang, T., Feng, G., Liang, J., & An, T. (2021). Acoustic scene classification based on Mel spectrogram decomposition and model merging. *Applied Acoustics, 182*, 108258.
5. Jothimani, S., & Premalatha, K. (2022). MFF-SAug: Multi feature fusion with spectrogram augmentation of speech emotion recognition using convolution neural network. *Chaos, Solitons & Fractals, 162*, 112512.
6. Boddapati, V., Petef, A., Rasmusson, J., & Lundberg, L. (2017). Classifying environmental sounds using image recognition networks. *Procedia Computer Science, 112*, 2048.
7. Salamon, J., & Bello, J. P. (2017). Deep convolutional neural networks and data augmentation for environmental sound classification. *IEEE Signal Processing Letters, 24*, 279.
8. Piczak, K. J. (2015). Environmental sound classification with convolutional neural networks. In *IEEE 25th International Workshop on Machine Learning for Signal Processing*.

Part III
Modern Electronic Information System and Control Technology

Runtime Accuracy Tunable Approximate Floating-Point Multipliers

Younggyun Cho, Luke Yin, and Mi Lu

1 Introduction

"The effect of big data is being felt everywhere from business to science and from government to arts" [1]. To extract valuable information from big data, we should properly deal with extremely large and complex data, which is very difficult to process using conventional arithmetic units due to a lack of computational power. Applications such as machine learning [2], image and signal processing [3], computer vision, and big data analytics [4] are all data-driven. For such applications, modern systems are striving to meet their computational demands within available resources. To break through this challenge, one of the well-known methods is approximate computing. Since many applications are intrinsically error-resilient [5, 6], as long as the error tolerance margin of target applications is guaranteed, we can apply the approximate computing technique to arithmetic units to obtain high performance and high energy efficiency without causing chaos. Due to the fact that most of the energy consumption is caused by arithmetic operations [7], and the multiplier is the most expensive unit among arithmetic units in terms of area, delay, and power [5], in this paper, we propose a new approach to designing approximate floating-point multipliers called RATAMs (*R*untime *A*ccuracy *T*unable *A*pproximate Floating-point *M*ultipliers) for error-tolerance applications.

To design the proper approximate multipliers, we should have guarantees on an error rate, which is a fundamental challenge of approximate computing unit

Y. Cho (✉) · M. Lu
Texas A&M University, WEB, College Station, TX, USA
e-mail: yc2704@tamu.edu; mlu@ece.tamu.edu

L. Yin
University of Texas, EER, Speedway, Austin, TX, USA
e-mail: lukeyin25@utexas.edu

© The Author(s), under exclusive license to Springer Nature Switzerland AG 2024
S. Yeom (ed.), *4th International Conference on Electronics and Signal Processing*, Signals and Communication Technology,
https://doi.org/10.1007/978-3-031-36670-3_7

design [8]. Yan et al. [9] uses an error estimator relying on machine learning (ML) classifier to predict the detailed feature of upcoming input data but ML-based approximate arithmetic designs such as [5, 6, 9] has inevitable costs. Training ML classifiers consume extra computational resources, energy, and even time. Besides, for different applications, ML classifiers often have a re-training phase to increase their accuracy since the error rate from approximate computing highly depends on the training data. In this paper, to remove all extra costs caused by an error estimator based on ML classifiers, we propose a new approximate floating-point multiplier called RATAMs.

In particular, our main contributions to this paper are as follows:

1. Introduce RATAMs that change the level of accuracy by themselves at runtime.
2. Present approximate floating-point multipliers without an error estimator based on an ML classifier (no training and re-training phases required).

2 Preliminaries

Diverse approximate computing techniques have been studied from approximate logic circuits to approximate algorithms. Priya et al. [10–14] suggest multiplier designs for optimization. Cho and Lu [5, 6] propose are configurable approximate multiplier. It has a dual mode. For high performance, the reconfigurable approximate multiplier can run in parallel. Each sub-multiplier computes two input pairs at the same time so that the performance increases. On the other hand, in the high accuracy mode, two sub-multipliers compute a single input pair. Thus, the error rate decreases. The high accuracy mode is for better output quality. An error estimator determines the accuracy modes at runtime.

3 Proposed Approximate Multipliers

RATAMs are based on our previous design [5, 6]. Unlike [5, 6], RATAMs do not have any error estimator. First of all, to design RATAMs as illustrated in Fig. 1, we need a clue to determine how frequently RATAMs need to change the accuracy mode. This indicator can be found by the cost function, *CostFunc* [5, 6], which takes into consideration all significant factors such as area (a), delay (d), and power (p) in Eq. (1), where err incurs extra cost when the error cannot be neglected. We denote the low bound of error as "low" and the upper bound of error as "up."

$$\text{CostFunc} = \begin{cases} a \times d \times p, & \text{err} \leq \text{low} \\ (1 + \text{err}) \times a \times d \times p, & \text{low} < \text{err} < \text{up} \\ \infty, & \text{err} \geq \text{up} \end{cases} \quad (1)$$

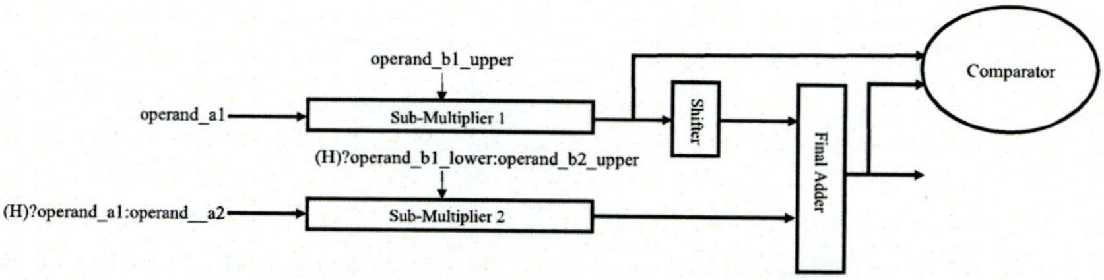

Fig. 1 RATAM; when RATAM with $k = 3$ and $k = 7$ is implemented, 3 bits or 7 bits in the mantissa field are utilized depending on the accuracy mode; (H)?A:B means if H is true, A is selected and if H is false, B is selected

Fig. 2 The percentage of data categorized into the optimal k

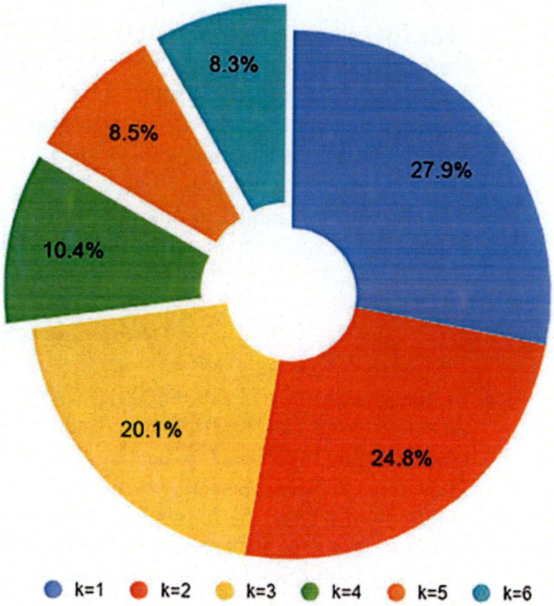

The given cost function, *CostFunc*, deals with the three different cases. First, the error rate is less than "low." We do not need to be concerned about the error rate since the error rate does not impact the quality of output. Second, the error rate is between low and up. We put a weight (1 + err) on $a \times d \times p$, and the *CostFunc* changes for various *err*. The level of accuracy k such that the *CostFunc* for which reaches the minimum is referred to as optimal k. Third, the error rate is out of the error tolerance margin (up) of target applications. We set the cost function to infinite so that the case can be removed from the choice of optimal k. Almost 28% (10.4% + 8.5% + 8.3%) of data require the high accuracy mode ($k > 3$), illustrated in Fig. 2.

Table 1 3-bit XOR-based LFSRs

Clock	q2	q1	q0	!(q2&q0)
1	1	0	0	1
2	1	1	0	1
3	1	1	1	0
4	0	1	1	1
5	1	0	1	0
6	0	1	0	1
7	0	0	1	1

3.1 *Linear-Feedback Shift Registers (LFSRs)*

LFSRs are common in both hardware and software implementations, beneficial to lots of applications requiring a fast generation of pseudorandom number sequences [15]. Nearly 28% of inputs require the high accuracy mode as illustrated in Fig. 2. Our design starts from the low accuracy mode. To obtain the minimum cost, we need to change the accuracy level from low to high with a 28% chance. Although the high degree of a polynomial generates the sequence more randomly, we choose the degree 3 polynomial to minimize the hardware overhead. We use a primitive polynomial to generate a maximum-length sequence. The characteristic polynomial is in Eq. (2).

$$P(x) = x^3 + x + 1 \tag{2}$$

The seed of our LFSRs is "001." Table 1 shows the random sequence of 3-bit XOR-based LFSRs. If q0 is selected as the active-low signal, we have three "0"s and four "1"s with 0 indicating the accuracy change from low to high and 1 indicating vice versa. Denoting the logic AND of q2 and q0 as (q2&q0), and letting! signify logical NOT, the active-low signal!(q2&q0) can have two "0"s and five "1"s. $1 \times 2 + 2 \times 5 = 12$, hence 2/12 (16%) data can select the high accuracy mode. For our experiments, we exploit both control signals, q0 and !(q0&q2), to trade off between error rates and runtime.

3.2 *RATAMs*

RATAMs, illustrated in Fig. 1, have some similarities to our previous work [5, 6], which relies on an error estimator based on an ML classifier. The low and high accuracy modes are supported. The shifter is used to adjust the alignment to add up the results of two sub-multipliers in the high accuracy mode. RATAMs utilize a comparator to check whether or not any difference exists in results between the low and high accuracy modes. In the high accuracy mode, the output of the sub-multiplier before the shifter is the same as the result for the low accuracy mode, and the final output of the high accuracy mode is the one after the full adder. To design a cost-efficient operator, the comparator can be implemented by various bit widths. From our experiments, we find out that the difference between the results from the low and high accuracy modes is detected by the least significant bit in the exponent field and several bits in the mantissa field. Let mant_1 and mant_2 be the mantissa fields of the high and low accuracy modes, respectively. The similarity holds for the exponent. [0] indicates the least significant bit in a field, while [s:t] indicates several selected bits, from bit s to bit t. The comparator is to compare whether exp_1[0] = exp_2[0] and man_1[22:t] = man_2[22:t]. t is to be determined, or the appropriate number of bits to be compared in the mantissa field is to be found by Algorithm 1.

Algorithm 1 Selecting a RATAM intellectual property (IP) and an index for a comparator

Input: application; the maximum error tolerance of the application (*Error*); a set of RATAMs of size N (IP[0:$N-1$]); an array to store t (n_comp[0:$N-1$]).
Output: RATAM IP[i]; t.

```
i = 0;
while i < N do
    for j = 1, 2, ..., 23 do        /* j is the number of bits to retain */
        t = 23 − j;
        Run the application with IP[i] using a comparator to compare mant_1[22:t]
        and mant_2[22:t];           /* After running the application, the approximate
                                       results are generated */
        Compute error_rate in the application;
        if (error_rate ≤ Error) then
            n_comp[i] = t;          /* store t in the array */
            break;
        end if
    end for
    if (n_comp[i] is NULL) then
        discard IP[i] and n_comp[i];
    end if
    i++;
end while
Search for index i based on the minimum runtime and energy consumption;
Return IP[i] and t such that n_comp[i] = t;
```

When the discrepancy is identified by our proposed comparator, we keep staying in the high accuracy mode. Otherwise, we determine to go back to the low accuracy mode for the next input pair. Figure 3a directly uses the control signal from 3-bit XOR-based LFSR. There is no comparator. We denote this to randFSM (random sequence from the LFSR). On the other hand, when the current state is S1, as

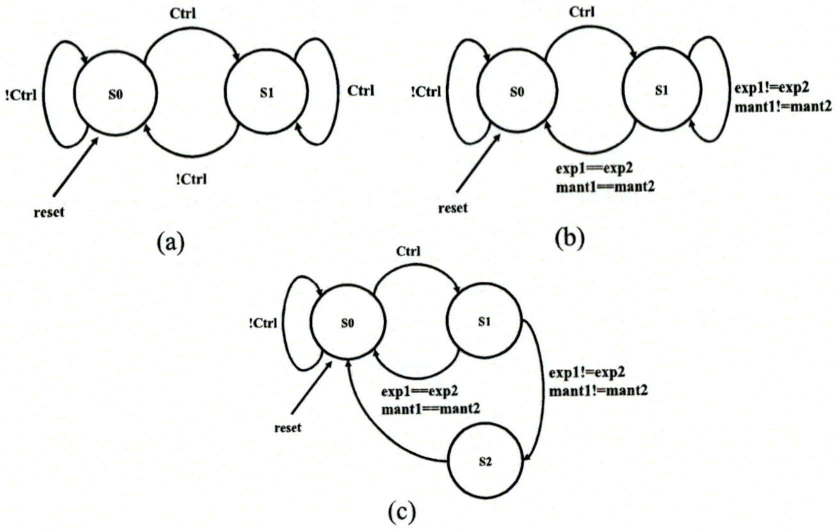

Fig. 3 (**a**) is random, (**b**) is to keep checking, and (**c**) is to check once, S0: the low accuracy mode S1, S2: the high accuracy mode, Ctrl: q0 or !(q2&q0) from LFSRs

illustrated in Fig. 3, a comparison of the exponent fields of the high and low accuracy modes starts being operated and same for the mantissa fields. If there is no difference detected, the next state is going back to S0, the low accuracy mode. Otherwise, the next state is the same as the current state to repeat the high accuracy mode for the next input pair. This FSM is named kcFSM (keep checking in S1). Similar to Fig. 3b, c operates the comparison once the current state is S1. However, to prevent the high accuracy mode from repeating too many times, this FSM conducts the comparison only once if the current state is S1. The next state changes from S1 to S2 if there is a discrepancy detected. Then, regardless of any condition, the next state changes from S2 to S0 (the next state returns back to the low accuracy mode). We indicate this FSM as coFSM (check once in S1). The details to select an RATAM and find the optimal number of bits for a comparator are presented in Algorithm 1.

We develop a new framework named RATAMs that has a self-checking control unit. This control unit helps change the level of accuracy at runtime. This framework relies on the fact that repeating the high accuracy mode is mandatory when the comparator detects the discrepancy. RATAMs intrinsically share and reuse sub-multipliers for the high-accuracy operations. There is no difficulty to obtain both results of the low and high accuracy modes. The hardware configuration of RATAMs is appropriate to integrate a comparator to check the discrepancy between results.

4 Experimental Results

4.1 RATAMs

To obtain the runtime and energy consumption of our proposed approximate multipliers, the input data set, 2000 floating-point data, is randomly generated. The input numbers are between 0 and 100 and follow a uniform distribution. We have three different types of RATAMs. They are differentiated by levels of accuracy ($k = 1$ & $k = 3$), ($k = 2$ & $k = 5$), and ($k = 3$ & $k = 7$). Each design is combined with three different FSMs in Fig. 3. The switching signal (Ctrl) of FSMs is from the proposed LFSRs, q0 and !(q2&q0). Initially, S0 switches to S1 according to the switching signal, and then, depending on the comparator, the next state becomes either the low accuracy mode or the high accuracy mode. As we discussed in Sect. 3, the number of bits in the mantissa field for comparison is important to determine the runtime and the error rate of RATAMs. Our experiments of the kcFSM and coFSM are conducted based on various bit-width for a comparator. To compare RATAMs with our previous designs [5, 6], referred to as Cho20, which is based on an ML classifier, the error rates, runtime, and energy consumptions are provided in Table 2.

The randFSM utilizes a pure pseudorandom sequence. Table 3 shows that when the Ctrl is q0, error rates decrease compared to !(q2&q0) since q0 has higher chances to be in the high accuracy mode, but runtime and energy consumption increase. The randFSM is a naïve approach to changing the levels of accuracy because the accuracy mode is entirely dependent on a pseudorandom sequence, not the current input pair.

The performance of a comparator-based design is associated with the number of bits in the mantissa field utilized. Table 4 provides error rates, runtime, and energy consumption of kcFSM with respect to the number of bits (#Bit). With q0 using more bits for a comparator reduces error rates and increases runtime. Design options are determined according to the error rates the target applications tolerate. For instance, if the error tolerance margin is required to be 7%, the RATAM, say

Table 2 Cho20

k	Error	Runtime	Energy
1&3	12.20%	5.32	0.819
2&5	5.04%	5.77	0.917
3&7	2.90%	5.41	0.876

Table 3 RATAMs with the randFSM

Ctrl	k	Error	Runtime	Energy
!(q2&q0)	1&3	14.0%	5.171	0.806
	2&5	7.30%	5.705	0.924
	3&7	3.80%	6.206	1.030
q0	1&3	14.0%	5.991	0.958
	2&5	6.60%	6.234	1.028
	3&7	3.40%	6.623	1.119

Table 4 RATAMs with the kcFSM and coFSM

		kcFSM			coFSM			
Ctrl	k	#Bit	Error	Runtime	Energy	Error	Runtime	Energy
!(q2&q0)	1&3	1	13%	5.43	0.88	13%	5.68	0.91
		2	11%	6.44	1.03	12%	5.94	0.96
	2&5	1	6.7%	5.40	0.89	6.7%	5.59	0.93
		2	6.2%	5.68	0.93	6.5%	5.70	0.95
		3	5.5%	6.50	1.05	6.2%	6.08	1.02
	3&7	1	3.6%	5.78	0.97	3.7%	5.78	0.97
		2	3.5%	5.97	0.99	3.5%	5.85	0.98
		3	3.0%	6.29	1.05	3.3%	6.16	1.04
		4	2.5%	6.99	1.18	3.0%	6.47	1.10
q0	1&3	1	11%	6.02	0.98	12%	5.83	0.95
		2	10%	6.59	1.05	11%	6.19	1.01
	2&5	1	5.9%	6.00	1.02	5.9%	6.07	1.02
		2	5.3%	6.51	1.07	5.6%	6.17	1.03
		3	4.4%	7.03	1.14	5.5%	6.24	1.05
	3&7	1	3.2%	6.31	1.06	3.2%	6.20	1.06
		2	3.0%	6.56	1.08	3.1%	6.25	1.07
		3	2.5%	7.13	1.19	2.9%	6.56	1.12

!(q2&q0) with ($k = 2$ & $k = 5$) leveraging a 1-bit comparison in the mantissa field, could be the best choice. On the other hand, if the required error margin changes to 4%, this design option does not meet the error rate requirement. In such case, the best design option is the RATAM ($k = 3$ & $k = 7$) using !(q2&q0) with a 1-bit comparison in the mantissa field. Furthermore, when the target application requires an even lower error tolerance margin, we can reduce the error rates by increasing the number of bits for comparison. Table 4 also provides the characteristics of RATAMs using the coFSM. The coFSM has the characteristic that stays in the high accuracy mode only twice (S1 and S2) maximumly. When the error tolerance of our target application is 7%, the proposed RATAM ($k = 2$ & $k = 5$) with the kcFSM (using !(q2&q0) and 1-bit comparison) is a good candidate, which is 6.4% faster and consumes 2.9% less energy than Cho20 ($k = 2$ & $k = 5$). When the error tolerance margin of target applications is 4%, the proposed RATAM ($k = 3$ & $k = 7$) with the kcFSM (using !(q2&q0) and 1-bit comparison) is the best option. Its runtime is 6.4% slower than the runtime (if considered only) of Cho20 ($k = 3$ & $k = 7$). However, it is not a fair comparison because we ignored the extra time required for data profiling, training, and re-training phases in our previous design. Table 5 provides comparisons of RATAMs and an accurate multiplier.

Table 5 Comparisons of an accurate multiplier with RATAMs

	Error	Runtime	Energy	Speedup	Energy save
Accurate	N/A	9.847	1.398	N/A	N/A
$k = 2$ & $k = 5$ + randFSMwith!(q2&q0)	7.30%	5.705	0.924	42.0%	33.9%
$k = 2$ & $k = 5$ + kcFSMwith!(q2&q0)	6.70%	5.400	0.890	45.2%	36.3%
$k = 3$ & $k = 7$ + coFSMwith!(q2&q0)	3.60%	5.785	0.971	41.2%	30.5%

Table 6 Experiment results for Fig. 4

	PSNR [dB]	SSIM	Runtime [ms]	Energy [mJ]	Speedup	Energy save
(b)	N/A	1	193.599	27.491	N/A	N/A
(c)	32.20	0.987	112.146	18.167	42.00%	33.91%
(d)	35.64	0.991	134.278	22.424	30.60%	18.43%
(e)	33.77	0.988	120.529	20.007	37.74%	27.72%
(f)	34.11	0.991	130.487	21.791	32.59%	20.73%
(g)	39.18	0.996	125.022	21.003	35.42%	23.60%
(h)	39.03	0.996	125.286	21.173	35.28%	22.98%

4.2 Image Processing

In image filter domains, lots of applications have a certain amount of tolerance for errors, and Gaussian filters are one of the popular image filters for the removal of the noises that an image has. When we use a Gaussian filter to process images, we need a two-dimensional function [16]. To apply the Gaussian filter to the image in Fig. 4a, the border values should be maintained, and the 5 by 5 filter coefficients become less while increasing the distance from the current pixel to the center of the filter. RATAMs are exploited to multiply the 25-pixel values by 25 weights. Figure 4a is the original, and Fig. 4b is the image after applying the Gaussian filter using the exact multiplier. Figure 4c is the image generated using RATAM ($k = 2$ & $k = 5$) with the randFSM. The difference between Fig. 4b, c is easy to see at a glance. Figure 4d, e, f are the resulted images using RATAM ($k = 2$ & $k = 5$) with the kcFSM and coFSM. As we can see, these images are much smoother than the one in Fig. 4c. However, if we carefully scrutinize these images by comparing them with that in Fig. 4b, we can find some discrepancies around the shoulder. Our RATAM ($k = 3$ & $k = 7$) with the kcFSM and coFSM are leveraged to create the images, Fig. 4g, h. This time, it is very hard to find any differences in Fig. 4b, g, h. Table 6 summarizes the results of the Gaussian filter using an exact multiplier as well as the proposed RATAMs. Here, quality metrics include the peak signal-to-noise ratio (PSNR) and structural similarity index measure (SSIM) values. (c) in Table 6 is 42% faster than (b) at the cost of smoothness, specifically for the shoulder part. (g) in Table 6 is 35.42% faster than (b). It is hard to find any discrepancy between Fig. 4b, g. According to the SSIM index, the difference is only 0.4%.

Fig. 4 Gaussian filter with different RATAMs, **(a)** original image, **(b)** using an exact multiplier, **(c)** RATAM ($k = 2$ & $k = 5$) with the randFSM (!(q2&q0)), **(d)** RATAM ($k = 2$ & $k = 5$) with the kcFSM (q0), **(e)** RATAM ($k = 2$ & $k = 5$) with the kcFSM (!(q2&q0)), **(f)** RATAM ($k = 2$ & $k = 5$) with the coFSM (q0)), **(g)** RATAM ($k = 3$ & $k = 7$) with the kcFSM (!(q2&q0)), **(h)** RATAM ($k = 3$ & $k = 7$) with the coFSM (!(q2&q0))

5 Conclusions

In this paper, we develop new approximate floating-point multipliers called RATAMs: *R*untime *A*ccuracy *T*unable *A*pproximate Floating-point *M*ultipliers. The key characteristic of RATAMs is that they can operate without an error estimator based on an ML classifier. Our previous design [5, 6] having an error estimator based on the kNN classifier requires inevitable data profiling, training, and re-training phases, which incur further computational resources, time, and energy. In contrast, RATAMs have a simple comparator on behalf of a complicated error estimator based on an ML classifier. Our simulation results show that RATAMs are able to improve performance and reduce energy consumption dramatically compared to an accurate multiplier. One of the RATAMs can achieve up to 45.2% speedup and 36.3% energy save with less than a 7% error rate. When the error tolerance of the target application is 4%, 41.2% speedup and 30.5% energy saving are accomplished by our proposed design. In the image filter domain, an RATAM can reduce the runtime and the energy consumption by 42% and 33.91%, respectively, with a 1.3% difference in the SSIM index.

References

1. Cukier, K. (2010). Data, data everywhere. *Economist, 394*(8671), 3–5.
2. Nazemi, M., Pasandi, G., & Pedram, M. Energy-efficient, low-latency realization of neural networks through boolean logic minimization. In *Proceedings of the 24th Asia and South Pacific Design Automation Conference 2019*, pp. 274–279.
3. Huang, J., Wang, B., Wang, W., & Sen, P. (2015). A surface approximation method for image and video correspondences. *IEEE Transactions on Image Processing, 24*(12), 5100–5113.
4. Sarwar, S. S., Srinivasan, G., Han, B., Wijesinghe, P., Jaiswal, A., Panda, P., Raghunathan, A., & Roy, K. (2018). Energy efficient neural computing: A study of cross-layer approximations. *IEEE Journal on Emerging and Selected Topics in Circuits and Systems, 8*(4), 796–809.
5. Cho, Y., & Lu, M. (2020). A reconfigurable approximate floating-point multiplier with kNN. In *2020 International SoC Design Conference (ISOCC) 2020*, pp. 117–118, IEEE.
6. Cho, Y., & Lu, M. (2022). Runtime accuracy alterable approximate floating-point multipliers. *Int Rob Auto Journal, 8*(2), 52–56.
7. Vahdat, S., Kamal, M., Afzali-Kusha, A., & Pedram, M. (2019). TOSAM: An energy-efficient truncation- and rounding-based scalable approximate multiplier. *IEEE Transactions on Very Large Scale Integration (VLSI) Systems, 27*(5), 1161–1173.
8. Miguel, J. S., & Jerger, N. E. (2016). The anytime automaton. *ACM SIGARCH Computer Architecture News, 44*(3), 545–557.
9. Yan, M., Song, Y., Feng, Y., Pasandi, G., Pedram, M., & Nazarian, S. (2019). KNN-CAM: A k-Nearest neighbors-based configurable approximate floating point multiplier. In *20th International Symposium on Quality Electronic Design (ISQED) 2019*, pp. 1–7, IEEE.
10. Priya, K. P., Kumar, P. S., Mounika, B., Lavanya, P., Das, D. S., & Babu, B. B. (2014). Review on optimization techniques of wallace tree multiplier. *International Journal of Electrical and Electronic Engineering & Telecommunications, 3*(2), 32–40.
11. Bala, K. P., & Padma, P. K. (2016). Implementation of power efficient 12-bit vedic multiplier by using ANT architecture. *International Journal of Electrical and Electronic Engineering & Telecommunications, 5*(4), 28–36.
12. Jalihal, V. B. (2014). Design and implementation of 32 bit unsigned multiplier using CLAA, CSLA, ETA. *International Journal of Electrical and Electronic Engineering & Telecommunications, 3*(3), 111–118.
13. Mohan, C. P., & Sumalatha, M. (2015). Realization of low power and area reduced aging-aware multiplier design. *International Journal of Electrical and Electronic Engineering & Telecommunications, 4*(4), 53–61.
14. Priya, D. S., & Krishna, K. H. (2014). Implementation of speed and power optimized vedic multipliers using reversible gate approach. *International Journal of Electrical and Electronic Engineering & Telecommunications, 3*(4), 107–117.
15. Linear-feedback shift register. https://en.wikipedia.org/wiki/Linearfeedback_shift_register. Last accessed 2022/12/02.
16. Gaussian filter. https://en.wikipedia.org/wiki/Gaussian_filter. Last accessed 2022/12/02.

Experimental and Numerical Analysis of the Heat Dissipation Characteristics of a Battery Cabin of an Electrical Bus

Qian Peng, Mengliang Si, Fenggang Han, and Huaijie Pan

1 Introduction

Coincident with the rapid development of electric buses, the reliability of batteries has become a substantial concern in the transport industry [1, 2]. Among the numerous related problems, thermal runaway constitutes a major issue for batteries during charging and discharging processes once the heat cannot be dissipated effectively [3]. Accordingly, the accumulation of heat in the battery cabin, which is packed with a number of batteries, will have a negative effect on the dissipation of heat away, even though each battery contains an individual cooling system.

Researchers have conducted multiple beneficial investigations into the heat dissipation characteristics of vehicle power sources. For example, Li Zhongxing et al. studied the rising temperatures of electric vehicles under different driving conditions [4]. Anosh Mevawalla et al. studied the law of heat release of lithium-ion batteries at different discharge rates through experiments and numerical simulation [5]. Sun Xudong et al. studied the thermal runaway characteristics of lithium-ion batteries under different heat dissipation conditions and different thermal stability materials through numerical simulation [6]. Furthermore, Zhou Haobing et al. studied the thermal behaviour of battery module and the flow field of air under different discharge rates through numerical simulation and analysed the effects of

Q. Peng (✉) · F. Han
School of Mechanical and Automotive Engineering, Xiamen University of Technology, Xiamen, China

Xiamen Vehicle Design & Services Co., Ltd., Xiamen, China

M. Si · H. Pan
School of Mechanical and Automotive Engineering, Xiamen University of Technology, Xiamen, China

© The Author(s), under exclusive license to Springer Nature Switzerland AG 2024

S. Yeom (ed.), *4th International Conference on Electronics and Signal Processing*, Signals and Communication Technology,
https://doi.org/10.1007/978-3-031-36670-3_8

orifice parameters, inlet pressure and discharge rate on the performance of air-cooling strategy [7]. Fan Yiwei et al. proposed a new double-layered dendritic channel liquid cooling system based on structural theory and carried out multi-objective optimization design [8]. Yao Chengning et al. proposed a novel micro heat pipe array (MHPA) for fast charging and special on-vehicle scenarios, which is used for thermal management in a wider research range [9]. Prashant Jindal et al. studied the heat dissipation effectiveness of the coolant of graphene nanoplatelets with different concentrations through numerical simulation [10]. The current research mainly focuses on the cooling and heat dissipation system of the battery, the coolant material, the battery of the electric vehicle, etc., while the research on the air-cooling system of the electric vehicle is less.

This paper aims to study the heat dissipation characteristics of the battery cabin of an electric bus and the temperature distribution of the battery cabin in a real driving cycle. Moreover, the heat dissipation characteristics of the battery cabin with different inlet and exhaust grilles are investigated.

2 Field Test

The facility used in this experiment is mainly composed of the following: a chassis dynamometer, a high-power wind tunnel fan and a FLUKE Ti400 thermal imager with the corresponding SmartView post-processing software. The thermal imager is shown in Fig. 1, which is a particularly important piece of test equipment used to obtain temperature information, a 5 million-grade industrial camera with a measuring temperature range of −20 °C~1200 °C and a heat sensitivity of ≤0.05 °C.

Fig. 1 FLUKE Ti400 thermal imager

Fig. 2 Tested bus

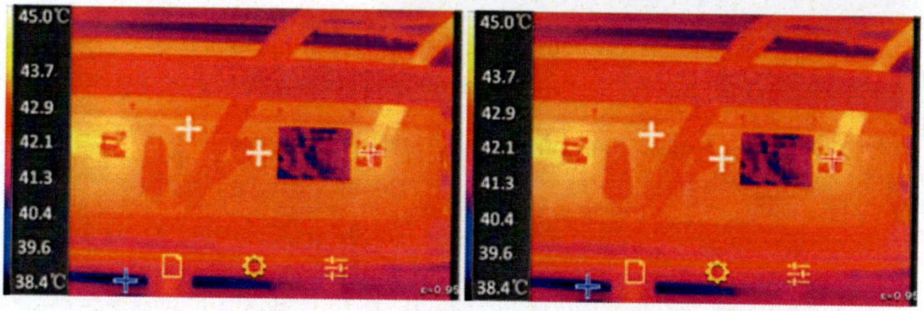

Fig. 3 Temperature distribution of the battery obtained through thermal imaging

Figure 2 shows a field test of 10.5 m electric city bus was carried out according to the characteristics of city bus in the summer climate of South China with the ambient temperature of 35 °C. The test was carried out at a commercial vehicle testing centre in Xiamen. The driving cycle of a bus was generated through a chassis dynamometer, and a high-power wind tunnel fan was placed in front of the chassis dynamometer to provide the airflow around the bus in the test. The bus was driven at a constant speed of 40 km/h, and the temperature of the battery cabin was measured by the thermal imager to obtain the thermal distribution. SmartView post-processing software was employed to observe the thermal characteristics with the thermal imager.

The maximum temperature and temperature difference in the battery cabin are used to evaluate the heat dissipation. The temperature distribution of the battery cabin obtained through infrared thermal imaging is shown in Fig. 3, in which the temperature is 38.4 °C~45 °C. The temperatures of the batteries are significantly higher than that of the cabin, and the temperatures around wires are also relatively higher resulted by the high electric current; in addition, the temperature of the frame is higher than that of the surrounding environment due to heat conduction of the steel frame.

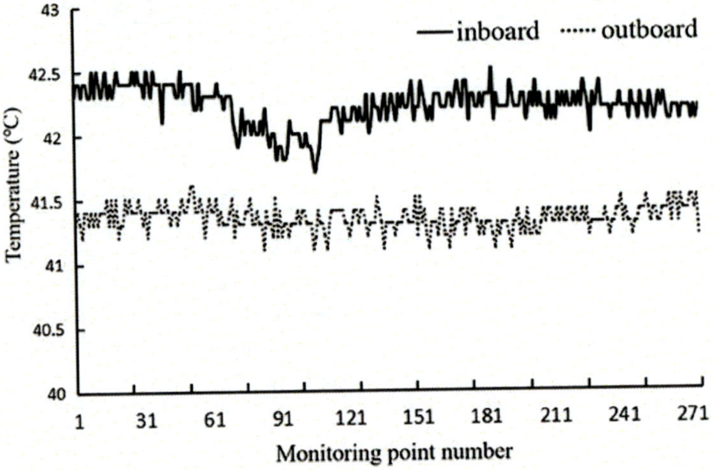

Fig. 4 Temperatures of the inboard and the outboard of the cabin

The temperature distributions of the inboard and the outboard of the battery cabin are shown in Fig. 4, which indicates that the temperature difference between the inboard and the outboard of the battery cabin is approximately 1~2 °C. With the continuous heat generation of batteries, the heat dissipated into the cabin cannot be dissipated immediately, whereas the outboard of the cabin can quickly dissipate heat because of the natural airflow in the presence of wind. As a result, the inboard temperature of the cabin is slightly higher than that of the outboard.

3 Modelling

3.1 Mathematical Model

Battery and their corresponding supporting structures are arranged in the battery cabin in the middle of the bus. The airflow within the cabin is turbulent, and the heat generated by the batteries is mainly transferred in the form of forced convective transfer and heat conduction between solids. The K-ε turbulence model is used to establish the mathematical model [11, 12], and the model includes the following equations.

Please note that the first paragraph of a section or subsection is not indented. The first paragraphs that follow a table, figure, equation, etc. does not have an indent, either.

1. Mass conservation equation, the expression for which is:

$$\frac{\partial \rho}{\partial t}+\frac{\partial\left(\rho v_r\right)}{\partial r}+\frac{\partial v_{\mathrm{r}}}{\partial r}+\frac{\partial \rho v_\phi}{r^{\partial \phi}}+\frac{\partial \rho v_z}{\partial z}=0 \tag{1}$$

where r (radial), z (axial) and φ (circular) are parameters in the cylindrical coordinate system; vr, vz and vφ are relative fluid velocities in the three directions; ρ denotes the gas density; and t represents time.

2. The momentum equation, which is expressed as follows (where the effect of gravity on the gas is ignored):

$$\frac{\partial(\rho V)}{\partial \mathrm{t}}+\nabla(\rho V \otimes V)-\nabla\left(\mu_{\mathrm{eff}} \nabla V\right)=\nabla_{\mathrm{p}}+\nabla\left(\mu_{\mathrm{eff}} \nabla V\right)^T \tag{2}$$

where V is the fluid velocity vector; μ_{eff} is the equivalent viscosity, $\mu_{\text{eff}} = \mu + \mu_t$, where μ is the fluid viscosity, μ_t is the eddy viscosity, $\mu_t = C_\mu \rho k^2/\varepsilon$, where C_μ is the constant 0.09; and P_t is the correction pressure, $P_{\mathrm{t}} = P + 2\rho k/3$, where P is the pressure.

3. The energy equation, which is expressed as

$$\frac{\partial\left(\rho h_t\right)}{\partial \mathrm{t}}-\frac{\partial p}{\partial t}+\nabla\left(\rho v h_t\right)=\nabla(\lambda \nabla T) \tag{3}$$

where h_t is the total enthalpy of gas, λ represents the thermal conductivity and T denotes the gas temperature.

4. The turbulent kinetic energy equation (k equation), which is expressed as

$$\frac{\partial(\rho k)}{\partial \mathrm{t}}+\nabla(\rho \mu k)=\nabla\left[\left(\mu+\frac{\mu_t}{\sigma_{\varepsilon}} \nabla k\right)+p_k-\rho \varepsilon\right] \tag{4}$$

5. The turbulence dissipation rate equation, which is expressed as

$$\frac{\partial(\rho \varepsilon)}{\partial \mathrm{t}}+\nabla(\rho \mu \varepsilon)=\nabla\left[\left(\mu+\frac{\mu_t}{\sigma_{\varepsilon}}\right) \nabla \varepsilon\right]+\frac{\varepsilon}{K}\left(C_{\varepsilon 1} p_k-C_{\varepsilon 2} \rho \varepsilon\right) \tag{5}$$

where $C_{\varepsilon 1}$, $C_{\varepsilon 2}$, σ_k and σ_ε are constants. According to the research results of Launder et al. [13] and subsequent related tests, the values of $C_{\varepsilon 1}$, $C_{\varepsilon 2}$, σ_k and σ_ε in the standard $K - \varepsilon$ model are $C_{\varepsilon 1} = 1.44$, $C_{\varepsilon 2} = 1.92$, $\sigma_k = 1.0$ and $\sigma_\varepsilon = 1.3$.

3.2 *Heat Generation Model for Lithium-Ion Batteries*

In this simulation, each cell is regarded as a separate enclosed subsystem that exchanges heat only with the exterior of the cell. The material of the cell is assumed to be the same. The specific heat capacity and thermal conductivity in the three directions, namely, X, Y and Z, are equal to the thermal conductivity of the battery and are not affected by the charge state or real-time temperature. The governing equation for the heat generation rate of the battery is as follows:

$$\dot{Q} = \sum_{i=1}^{n} \dot{q}_p + \sum_{i=1}^{n} \dot{q}_n \tag{6}$$

where $\dot{Q}$ is the total heat generated by the battery and n is the number of cells.

The heat generation of the battery in an electric vehicle in operation complies with the thermodynamic equilibrium equation expressed as

$$\dot{Q} = \dot{Q}_a + \dot{Q}_e \tag{7}$$

where $\dot{Q}$ is the heat absorption of the battery cell and $\dot{Q}_e$ is the heat exchange between the battery cabin and environment.

The heat absorption of the battery is calculated using the following equation:

$$Q_a = \sum_{i=1}^{n} \left(c_{pi} m_i \Delta T_i\right) \tag{8}$$

where C_{pi} is the average specific heat capacity of a single battery cell, J/(kg·K) is the average mass of the battery and ΔT_i denotes the temperature difference between batteries.

Generally, the mass and material of the cells in a battery are identical. Therefore, Eq. (9) can be simplified as follows:

$$Q_a = c_p m \sum_{i=1}^{n} \Delta T_i \tag{9}$$

The battery will generate substantial heat in a short time under limiting conditions. Accordingly, when the heat generation rate is higher than the heat dissipation rate, the temperature will continuously rise, and an excessive temperature will affect the cycle life and safety of the battery.

3.3 Battery Cabin of a Bus

The geometric modelling of the bus battery cabin is completed in the CREO software. The dimensions of the battery cabin are 1898 mm × 2340 mm × 897 mm. The power battery of an electric bus is equipped with a square lithium iron phosphate lithium-ion battery. A total of eight batteries are arranged in the middle of the battery cabin, and the upper and lower layers are superposed by four blocks. To guarantee

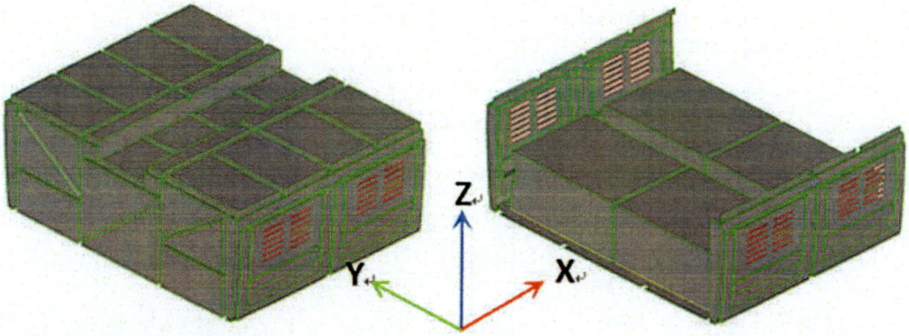

Fig. 5 Simplified geometric model of the battery cabin

the computational stability and efficiency of the resulting computational fluid dynamics (CFD) heat flow field, the geometric model is simplified appropriately according to the following principles:

1. The research mainly focuses on the inboard heat flow of the battery cabin.
2. The fluid computing domains should be closed. Therefore, the battery cabin in the middle section of the bus is closed, and the joints between parts in the cabin and the bus surface are ignored.
3. The upper and lower layers of the battery cabin are tightly arranged; thus the upper and lower battery cabins are regarded as a single entity. The simplified geometric model of the battery cabin in the middle section of the bus is shown in Fig. 5, where the X axis is parallel to the sidewall of the battery cabin and the Y axis is perpendicular to the sidewall of the battery cabin.

3.4 Research Scheme of the Inlet and Exhaust Grilles in the Battery Cabins of Electric Bus

As shown in Figs. 6, 7, and 8, three types of inlet and exhaust grilles employed in power battery cabin, namely, horizontal stripe grilles, vertical stripe grilles and circular-type grilles, are studied in this research. The cover of the battery cabin in an electric bus consists of four inlet planes, two of which are symmetrically distributed on each side. Take the horizontal stripe grille scheme as an example. Each inlet plane includes 6×2 air inlets, and each horizontal stripe grille is a 160 mm $\times$ 20 mm rectangle. The number of cross sections and their distribution and size for each of the three types of grilles are consistent. Figures 6, 7, and 8 illustrate the geometric models of the three types of grille schemes.

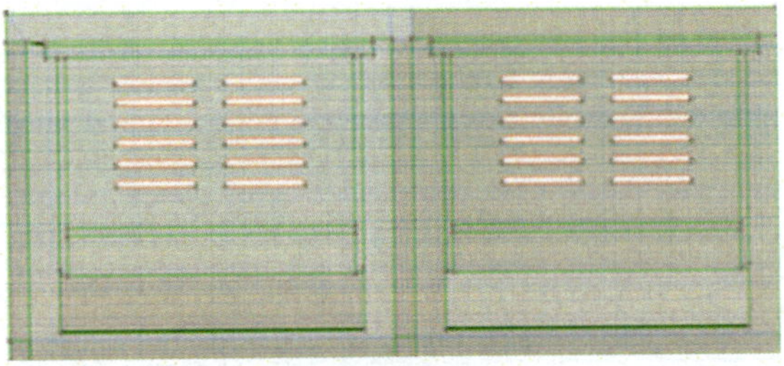

Fig. 6 Horizontal stripe inlet and exhaust grilles

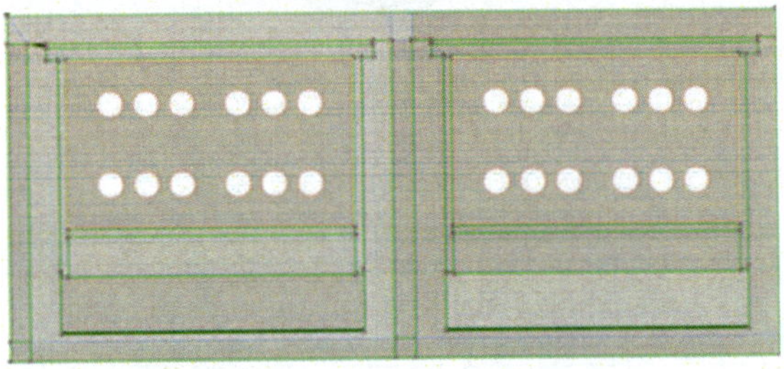

Fig. 7 Circular inlet and exhaust grilles

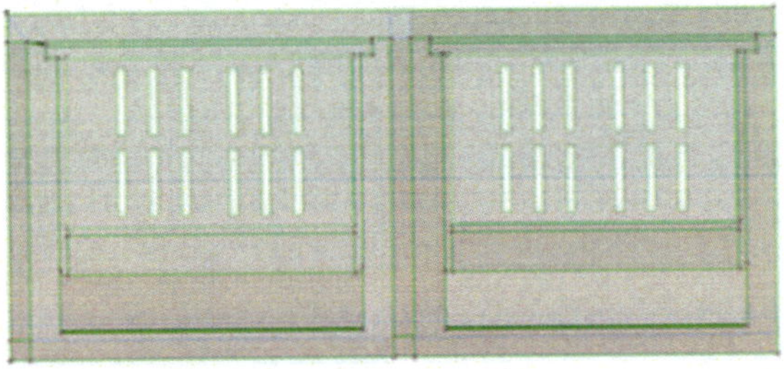

Fig. 8 Vertical stripe inlet and exhaust grilles

4 Thermal Analyses of the Battery Cabin

4.1 Finite Element Model

To simplify the calculation, four geometric cuboid models are constructed outside the inlet and exhaust grilles to simulate the inlet of natural air on both sides of the bus during driving cycle. The mesh network is divided into tetrahedral meshes, a total of 2,425,000 meshes are employed, and the fluid mesh is shown in Fig. 9.

4.2 Boundary Conditions

A velocity inlet and a pressure outlet are adopted as boundary conditions in the numerical simulation. The inboard of battery cabin is defined as the calculation domain range, while the outboard is excluded. The air vents on either side of the

Fig. 9 Cross bar fluid grid

battery cabin are set at a constant pressure, and the static pressure is 0 (where the reference pressure is 101,325 Pa), thereby allowing free airflow. A wall boundary considering solid heat transfer through the joint between the bottom of the battery cabin and the battery cabin itself is adopted for the battery cabin wall. The battery cabin and the external air exchange energy through convection and radiation heat transfer. According to the literature [14], the convective heat transfer coefficient hx = 15 W/(m^2·K), and the radiation heat transfer coefficient hr = 0.2.

Using FLUENT software to simulate the condition of buses working in the high temperature, setting the environment temperature at 308.15K. Considering the real driving cycle of city buses in China, two scenarios are considered: a uniform speed of 40 km/h and a uniform speed of 10.8 km/h on a 4% slope. The second-order upwind scheme is used for the spatial discretization of the governing equation, and the semi-implicit method for pressure-linked equations (SIMPLE) algorithm is used for the numerical solution. To minimize the influence of the mesh size, adaptive mesh refinement is carried out in areas with complex velocity and temperature gradients [15].

4.3 *Flow Field Analysis*

The flow characteristics of the battery cabin under the low-speed driving condition (10.8 km/h) on a slope are shown in Fig. 10. The X-Y cross section is located at the centre of the inlet and exhaust grilles at the top of the battery cabin. The maximum air velocities in the battery cabin corresponding to the three schemes consisting of a horizontal stripe grille, a vertical stripe grille and a circular grille are 0.78 m/s,

Fig. 10 Streamline distributions through the X-Y cross section in the middle of the battery cabin (10.8 km/h). (**a**) Horizontal stripe grille. (**b**) Circular grille. (**c**) Vertical stripe grille

Fig. 11 Streamline distributions through the X-Y cross section in the middle of the battery cabin (40 km/h). (**a**) Horizontal stripe grille (**b**) Circular grille (**c**) Circular grille

0.91 m/s and 0.06 m/s, respectively. The circular scheme demonstrates the largest airflow velocity, while the airflow velocity of the vertical stripe scheme is much lower than those of the other two schemes.

The streamline characteristics along the cross sections of the three groups of battery cabin show great differences. The horizontal stripe grille scheme shows eight regular and symmetrical air swirls of similar size, and an obvious horizontal airflow zone occurs in both the inlet and the outlet. The circular scheme also presents a clear horizontal airflow zone that passes transversely through both the inlet and the outlet in comparison with the horizontal stripe grille scheme. In contrast, the vertical stripe scheme presents only three disordered air swirls. In summary, the inlets and outlets in the circular and horizontal stripe grille schemes can better guide the flow of air and contribute to the efficient dissipation of heat in the cabin.

At a constant speed of 40 km/h, the streamline distributions of the battery cabin for the three schemes are shown in Fig. 11. The air velocities in the horizontal, circular and vertical grille schemes are 3.14 m/s, 5.19 m/s and 0.37 m/s, respectively. The overall airflow characteristics are similar to those under low-speed driving conditions on the slope. The velocity of the circular scheme is the highest, while that of the vertical stripe scheme is the lowest. Furthermore, the distributions of air swirls are similar to those under low-speed driving conditions on the slope.

To further explain the differences in the flow characteristics between the front and the rear grilles, two columns of inlet and exhaust grilles are studied, and the variations in the airflow through the Y-Z plane are described, as shown in Fig. 12.

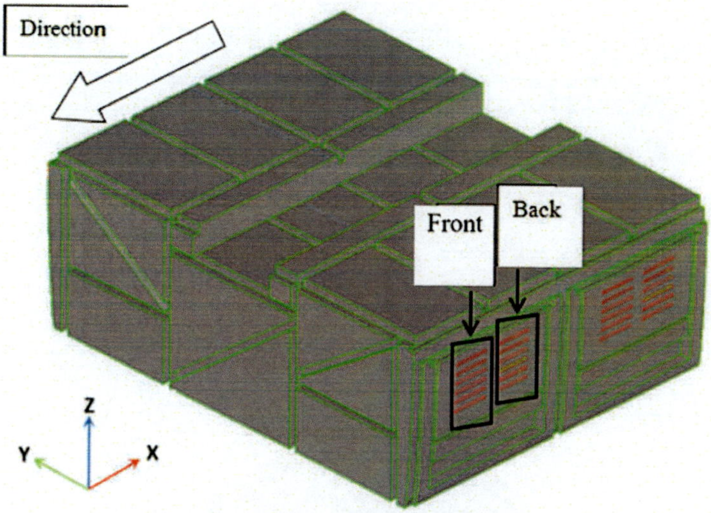

Fig. 12 Schematic diagram of the Y-Z plane section

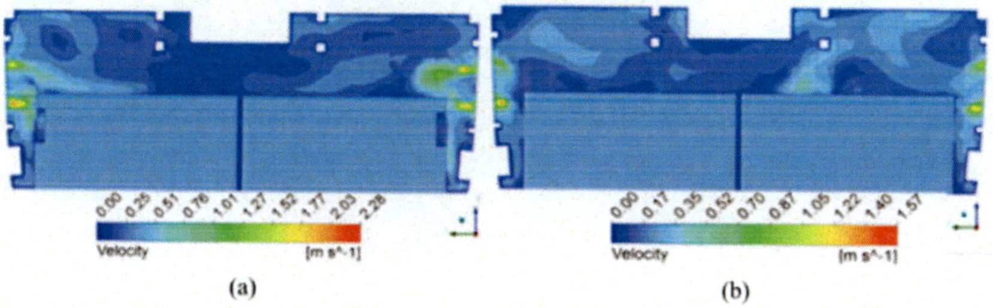

Fig. 13 Velocity contour through the Y-Z cross section in the circular scheme. (**a**) Front section (**b**) Rear section

Taking the flow field in the battery cabin at a speed of 40 km/h as an example, the air velocity characteristics through the Y-Z plane in the abovementioned three schemes are studied. As shown in Figs. 13 and 14, the airflow velocities in the circular scheme are 2.28 m/s and 1.57 m/s, respectively. Moreover, the airflow velocities at the inlet and outlet are the highest. Meanwhile, the air velocities in the front and back rows of the vertical stripe scheme are 1.02 m/s and 0.63 m/s, respectively, and the airflow velocities in the cabin are relatively uniform. In both schemes, the airflow velocity through the front grille section is higher than that through the rear grille section.

To reveal the variation law of the flow field in the battery cabin, two air velocity monitoring lines, each of which has 50 monitoring points with the same spacing, are set in the battery cabin in the middle of the bus. The locations of the monitoring lines are shown in Fig. 15.

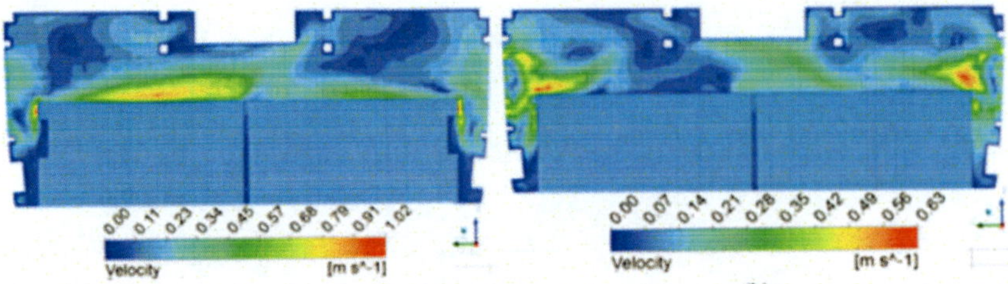

(a) (b)

Fig. 14 Velocity contour through the Y-Z cross section in the horizontal stripe scheme. (**a**) Front section (**b**) Rear section

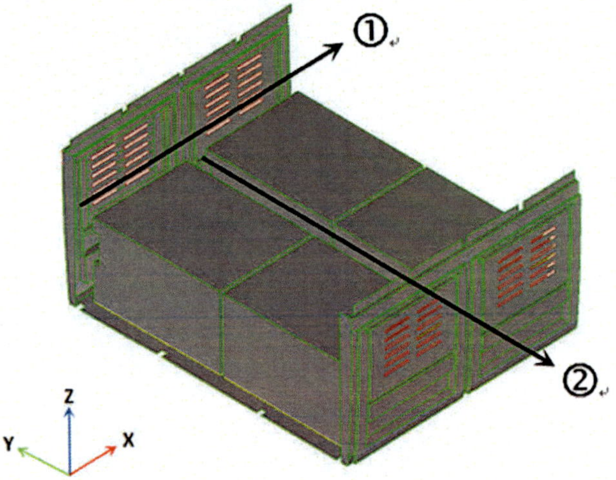

Fig. 15 Schematic diagram of the two monitoring lines

The variations in the air velocity in the No. 1 monitoring line in the battery cabin under the two scenarios are shown in Figs. 16a and 17a. The airflow shows the lowest variation in the vertical stripe grille scheme. The overall velocity in the horizontal stripe grille scheme is relatively high, particularly at the position closer to the cabin centre. The overall velocity in the circular scheme is the highest, and the velocities on the two sides of the cabin are symmetrical, while the velocity at the centre of the cabin is the lowest.

The variations in the air velocity in the No. 2 monitoring line (perpendicular to the No. 1 line) in the battery cabin under the two scenarios are shown in Figs. 16b and 17b. The air velocity near the sidewall of the cabin is higher than that at the centre of the cabin in the Y direction; in addition, the variation in the airflow velocity at the centre of the cabin is more complex.

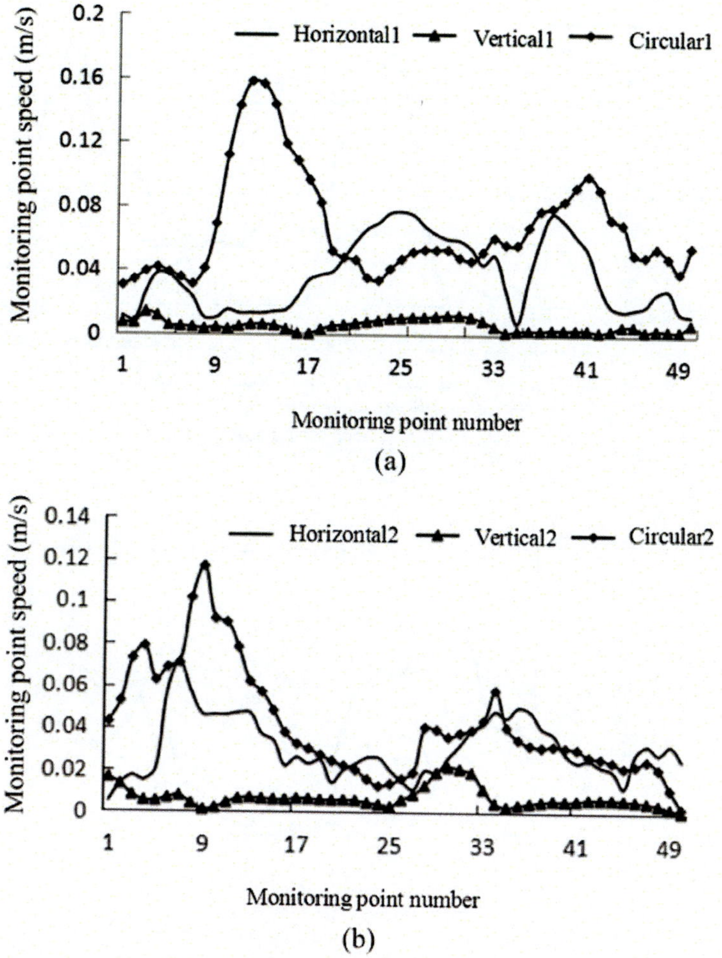

Fig. 16 Velocity curves along the monitoring direction at $V = 10.8$ km/h. (**a**) Changes in the air velocity along the No. 1 line. (**b**) Changes in the air velocity along the No. 2 line

4.4 Temperature Analysis

X-Y cross sections of the battery cabin temperatures under the three different grid schemes are shown in Fig. 18. The temperature within the channel between the two battery cabins in the Y direction is high, thereby forming a local high-temperature zone in the cabin in sharp contrast to the ambient temperature within the cabin. This indicates that the airflow through the gap is insufficient in all three types of grille inlets. Moreover, there is an air circulation channel in the central area of the

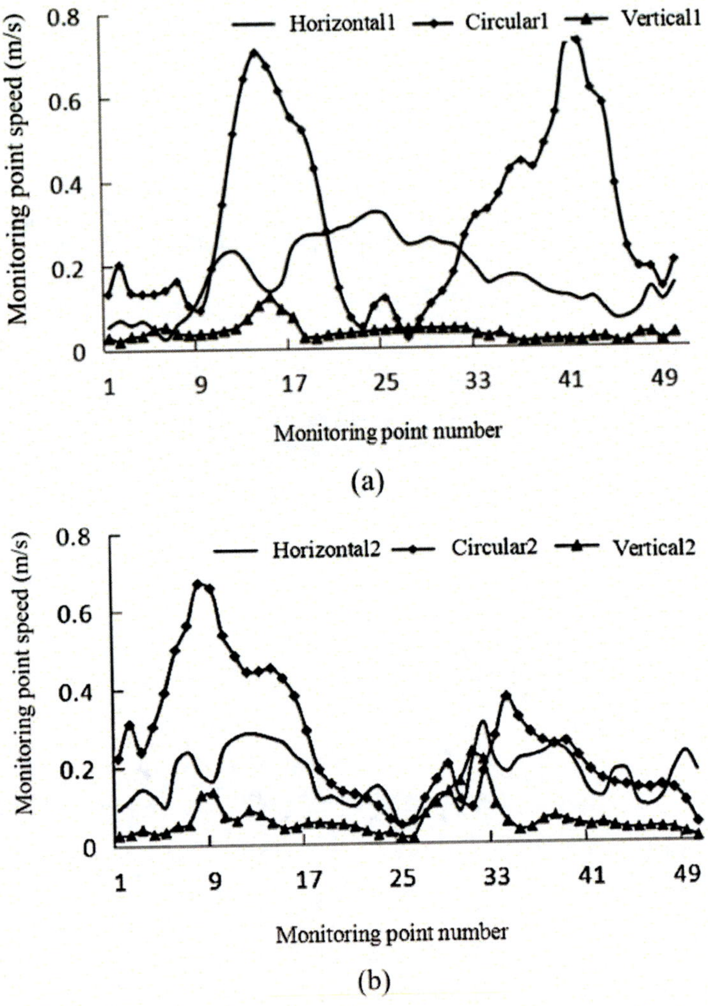

Fig. 17 Velocity curves along the monitoring direction at $V = 10.8$ km/h. (**a**) Changes in the air velocity along the No. 1 line. (**b**) Changes in the air velocity along the No. 2 line

Y direction and the air inlet area, leading to the maximum airflow and velocity; therefore, the temperature in this area is the lowest.

Figure 19 shows the distributions of the temperature throughout the cabin in the three different grid schemes under the low-speed driving condition (10.8 km/h) on the slope at high power. The temperature ranges of the cabin in the vertical, circular and horizontal stripe grille schemes are 313.8 K~321.3 K, 312.7 K~318.6 K and 314.4 K~326.3 K, respectively. The temperature in the vertical stripe grille scheme

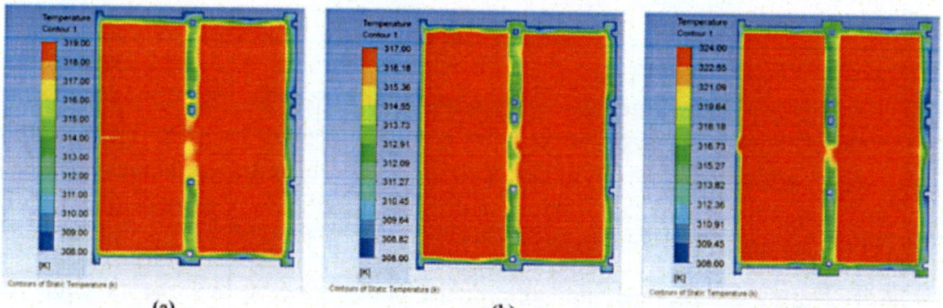

Fig. 18 X-Y cross sections of the temperatures at $h = 0.155$ m. (**a**) Horizontal stripe grille scheme (**b**) Circular grille scheme (**c**) Vertical stripe grille scheme

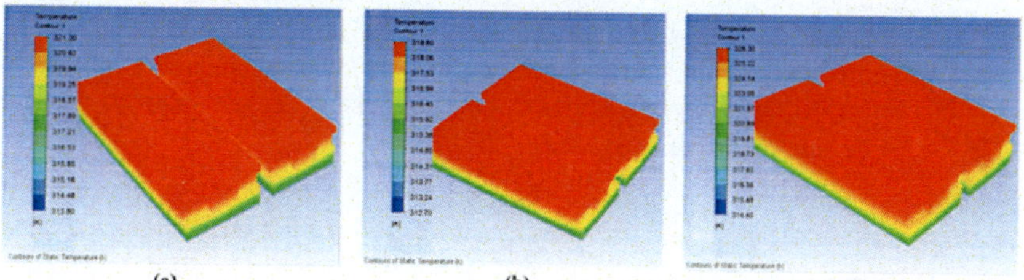

Fig. 19 Temperature distributions in the cabin. (**a**) Horizontal stripe grille (**b**) Circular grille (**c**) Vertical stripe grille

Table 1 Comparison of the simulation results of the flow field and temperature field

Item		Horizontal stripe grille	Circular grille	Vertical stripe grille
Maximum flow velocity (m/s)	10.8 km/h	0.84	1.2	0.13
	40 km/h	4.07	5.52	0.62
H = 0.155 m X-Y plane (m/s)	10.8 km/h	0.78	0.91	0.06
	40 km/h	3.14	5.19	0.37
Surface temperature of the cabin (K)	40 km/h	313.8–321.3	312.7–318.6	314.4–326.3

is the highest, which is consistent with the heat dissipation characteristics of the cabin. Table 1 summarizes the simulation results of the flow field and temperature field for three different grid schemes. Overall, the cabin in the circular scheme presents the optimal cooling effect, followed by the horizontal stripe scheme, while the vertical stripe scheme is the least ideal.

5 Conclusions

This paper studied the heat dissipation characteristics of the battery cabin of a pure electric bus using experimentation and numerical analysis. With regard to the design of both the inlet and the outlet of the battery cabin, three schemes consisting of a vertical stripe, horizontal stripe and circular grille were proposed. The following conclusions can be obtained:

1. During the real driving cycle of the purely electric bus, the air temperature in the cabin greatly increases, and when the bus runs at 40 km/h, the temperature difference in the battery cabin reaches 8 °C.
2. The circular and horizontal grille schemes can guide the transverse airflow through the inlet and outlet, which would be helpful for achieving effective heat dissipation within the battery cabin.
3. The battery cabin with a circular grille scheme achieves the optimal cooling effect, followed by the horizontal stripe grille scheme, while the vertical stripe grille scheme is the least ideal.

References

1. Sievers, M., Sievers, U., & Mao, S. S. (2010). Thermal modelling of new Li-ion cell design modifications. *Forschung im Ingenieurwesen, 74*(4), 215–231.
2. Chen, S. C., Wan, C. C., & Wang, Y. Y. (2005). Thermal analysis of lithium-ion batteries. *Journal of Power Sources, 140*(1), 111–124.
3. Feng, X., Zheng, S., Ren, D., He, X., Wang, L., Cui, H., & Ouyang, M. (2019). Investigating the thermal runaway mechanisms of lithium-ion batteries based on thermal analysis database. *Applied Energy, 246*, 53–64.
4. Zhongxing, L. I. (2014). Temperature study of pure electric vehicles battery pack at different driving conditions. *Journal of Mechanical Engineering, 50*(16), 180–185.
5. Mevawalla, A., Panchal, S., Tran, M. K., Fowler, M., & Fraser, R. (2020). Mathematical heat transfer modeling and experimental validation of lithium-ion battery considering: tab and surface temperature, separator, electrolyte resistance, anode-cathode irreversible and reversible heat. *Batteries, 6*(4), 61.
6. Sun, X., Xu, X., Zhao, L., Jin, H., Cao, L., & Wang, H. (2021). Blocking analysis of thermal runaway of a lithium-ion battery under local high temperature based on the material stability and heat dissipation coefficient. *Ionics, 27*(1), 107–122.
7. Zhou, H., Zhou, F., Xu, L., & Kong, J. (2019). Thermal performance of cylindrical Lithium-ion battery thermal management system based on air distribution pipe. *International Journal of Heat and Mass Transfer, 131*, 984–998.
8. Fan, Y., Wang, Z., & Fu, T. (2021). Multi-objective optimization design of lithium-ion battery liquid cooling plate with double-layered dendritic channels. *Applied Thermal Engineering, 199*, 117541.
9. Yao, C., Dan, D., Zhang, Y., Wang, Y., Qian, Y., Yan, Y., & Zhuge, W. (2020). Thermal performance of a micro heat pipe array for battery thermal management under special vehicle-operating conditions. *Automotive Innovation, 3*(4), 317–327.

10. Jindal, P., Sharma, P., Kundu, M., Singh, S., Shukla, D. K., Pawar, V. J., et al. (2022). Computational Fluid Dynamics (CFD) analysis of Graphene Nanoplatelets for the cooling of a multiple tier Li-ion battery pack. *Thermal Science and Engineering Progress, 31*, 101282.
11. Wang, F. (2004). *Computational fluid dynamics analysis: Principles and applications of CFD software*. Tsinghua University Press.
12. A ma no, R. S. (1984). Development of a turbulence near-wall model and its application to separated and reattached flows. *Numerical Heat Transfer, 7*(1), 59–75.
13. Launder, B. E., & Samaraweera, D. S. A. (1979). Application of a second-moment turbulence closure to heat and mass transport in thin shear flows—I. Two-dimensional transport. *International Journal of Heat and Mass Transfer, 22*(12), 1631–1643.
14. Feng, J. Y., Dai, Z. Q., Zhang, J. P., & Zhang, T. Z. (2013). Research on the Temperature Field Characteristics of LiFePO (4) Power Battery Based on Ansys Workbench 12. 0. Journal of Qingdao University (Engineering & Technology edition), 28(1).
15. Lan, K. T., & Srinivasan, K. (2009). *Influences of free stream conditions on vehicle thermal management–An analytical study* (No. 2009-01-1152). SAE Technical Paper.

The Design of an Automatic Proportional Electronic Flow Control System for Applying Agricultural Pesticides

Jhordan Deivis Surichaqui Oré, Camilo Raymundo Quesada, and Deyby Huamanchahua

1 Introduction

Farmers are exposed to various dangers when cultivating and trying to protect their crops from pests by using agrochemicals and organophosphates [1], which according to the World Health Organization (WHO) potentially cause acute or chronic toxic effects [2]. The indiscriminate use of these pesticides can have serious negative effects on the nervous, immune and reproductive systems, even causing cancer [3], and these affect not only humans but also nonhuman animals [4, 5]. In Peru, according to the Ministry of Health (MINSA), 82.1% of the cases of acute pesticide poisoning (IAP) are concentrated in six departments: Among them, in 2018 and 2019 Lima, Arequipa and Junín were the regions with highest number of reported cases of IAP [6].

In recent years, new technologies have been developed in precision agriculture to increase its production and to safeguard the health of farmers and their crops [4, 7]. This is a crucial development because agriculture could prove to be the key to reducing poverty for 80% of the world's low-income population [8]. For decades, one of the most effective methods to deal with pests has been to apply pesticides by using spray backpacks [9], but these require direct manipulation by an operator and therefore expose the operator and those around them to these chemicals [10]. Now, selective fumigation has become one of the most suitable methods to optimize the use of agricultural pesticides.

Faced with this problem of chemical exposure, different solutions have been proposed. One of them was the development of mobile robots, which would

J. D. S. Oré (✉) · C. R. Quesada · D. Huamanchahua
Universidad Continental, Huancayo, NJ, Peru
e-mail: 48042400@continental.edu.pe; 72575766@continental.edu.pe; dhuamanchahua@continental.edu.pe

© The Author(s), under exclusive license to Springer Nature Switzerland AG 2024

S. Yeom (ed.), *4th International Conference on Electronics and Signal Processing*, Signals and Communication Technology,
https://doi.org/10.1007/978-3-031-36670-3_9

apply the chemical substances through a remote control system [11]. One of the disadvantages of using these robots is the loss of precision in the application, which also wastes pesticides, because no system can yet apply the exact dosage that each plant needs. Furthermore, in another study [12], researchers developed flow control through a proportional–integral–derivative (PID) controller, for which they used the Lyapunov stability theorem to achieve an optimal output flow rate, which was mathematically modeled in Matlab to obtain the minimum profits for the integrator.

A precise spraying system that took into account the type of plant, the height of the plant and the separation distance between the plant and the machine was proposed by [13], who also vertically place opening nozzles according to levels required for optimal fumigation, obtaining 78% effectiveness. The optimization of the opening of the nozzle at the time of spraying the liquid was studied by [14], who measured the size of the drops by using an image-acquisition system, matching it to the nozzle pressure level, thus producing a table for spraying liquid pesticides.

To solve the problems raised above, this study aims to design a proportional flow control system for the automatic application of agrochemicals, which will be coupled with a prototype of a remotely guided mobile vehicle (robot) capable of moving between crops and dose them with agrochemicals in a selective and semi-autonomous way, thus optimizing the use of agricultural pesticides and reducing direct contact between farmers and pesticides during spraying. Finally, another of the objectives of this study is for it to serve as inspiration and support for the future development of a functional prototype that works in conjunction with a remote control system, which is planned to start after the publication of this study.

2 Methodology and Methods

For the design of the system, the guidelines of the VDI-2206 method were taken as a model, developed by the Association of German Engineers (Verein Deutscher Ingenieure, VDI) (VDI-RICHTLINIEN, 2004). This is a generalized and standardized procedure for design development [15], which is composed mainly of the fundamental stages for the development of mechatronic designs [16]. The mechanical, structural, electronic and control designs were derived from this method.

2.1 Definition of Requirements

To meet design expectations, a qualitative analysis of the salient details for the system design was developed. In this analysis, the environment to which the system will be exposed—in this case, agricultural land—was taken into account. For this, a machine must be able to execute all the processes starting from storage to dosing plants with agrochemicals. The machine size should not exceed 170 cm in width, because each row of corn crops is 75 cm between furrow and furrow; also, it must reach a height no less than 20 cm from the ground, and said height must be adjustable to the size of the crop [17]. Its storage capacity must be 20 liters; its structure must withstand corrosion and mechanical stress; and the system must be able to be configured manually and automatically.

2.2 Concept Generation and Evaluation

For the mobile equipment model that adapts to the needs of the sprinkler system, a mobile robot with simple radial wheels and a bolt fixing clamp was chosen; servomotors were selected to allow the tires to move; and the model was given an amplitude selective damper (ASD) and an adjustable arm with wide-angle conical nozzles for irrigation [18, 19]. An automatic electronic control system was equipped with infrared range sensors to calculate the height of plants, a variable flow electric pump, electronic control valves and a menu interface remotely controlled by the user.

2.3 Mobile Structure Design

The development of this system was inspired by needing a design with a structure capable of being mobilized by using a teleoperated robotic system in conjunction with a flow control system, which will be discussed later. Said design featured the dimensions required for its free movement across agricultural land, as shown in Fig. 1.

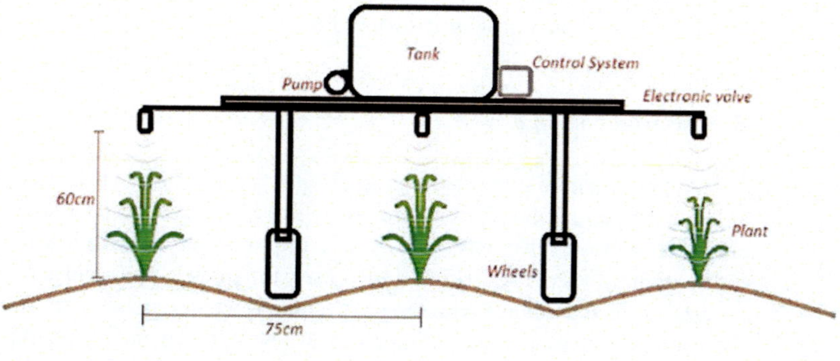

Fig. 1 Preliminary design of the mobile structure

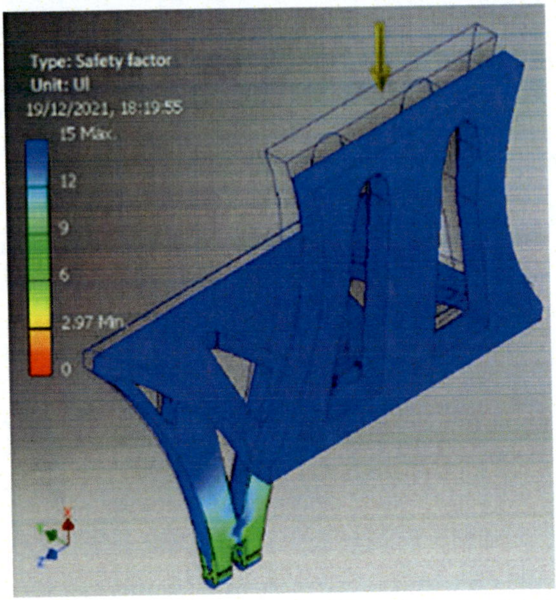

Fig. 2 Support metal structure simulation

2.3.1 Basic Form of the Design

Lunar and planetary rovers, such as the Perseverance or Sojourner, are effective when moving over rough terrain because their designs prioritize structural stability.

Both started from the basic geometric model for optimal load distribution, from which a wide range of designs have been obtained, which have been used as reference points for the development of a mobile structure, as shown in Fig. 2, where the structure can be seen being subjected to loads of 1030.05 N.

The support structure of the pesticide tank, motor and other accessories can distribute its weight on a grill, as shown in Fig. 3.

2.3.2 Electronic Design

Figure 4 shows a simulation of the proportional sprinkler system that is governed by an ArduinoMega 2560 microcontroller, which processes the data [20] provided by the infrared range sensors (GP2Y0A710), the same sensors that are also used to open the solenoid valves (EP015R + MP) used for spraying pesticides through the spray nozzles, in accordance with the data. Three reference points may be assigned by the user, which vary within a range of 10–60 cm, depending on the average minimum and maximum height of the plants.

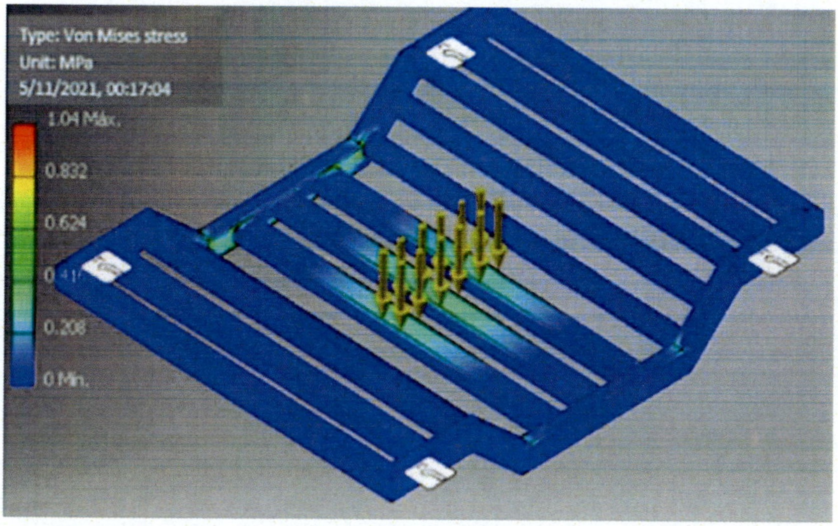

Fig. 3 Support base

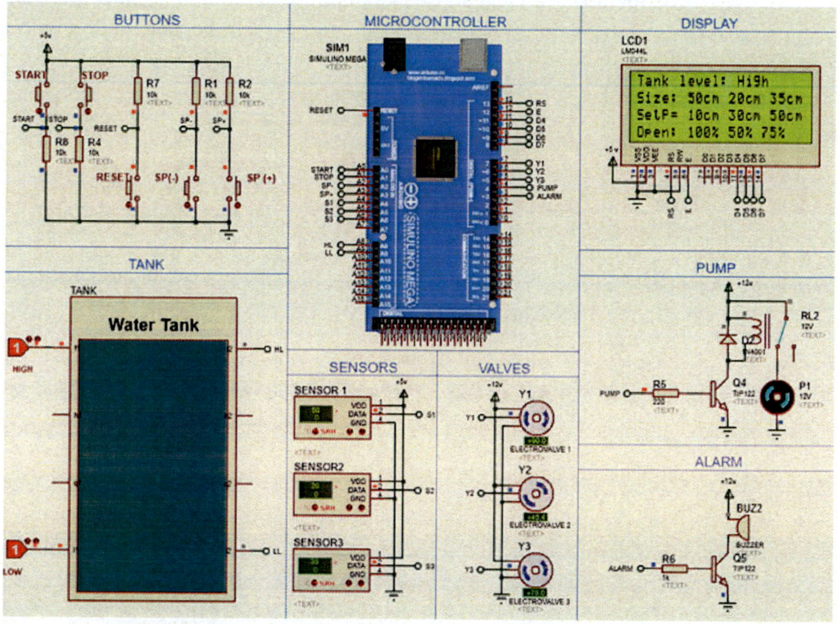

Fig. 4 Proportional sprinkler system simulation

2.3.3 Control Strategy

This design uses proportional control because it provides a fast response time and enables dosages customized to the size of each plant. The data on response times and dosages are collected by the sensors.

The system is based on Bernoulli's mathematical modeling (1), which takes into account the losses from materials and gains from external elements, such as the electric pump with an efficiency value, n, of 75%. With this, we can take advantage of the pressure within the system and use it to our advantage to reduce the consumption of electrical energy and compensate for the flow required for the spray nozzles.

$$Z_1 + P_{atm} + \frac{P_1}{\gamma} + \frac{V_1^2}{2g} + \frac{\mathrm{Pot}.n}{\gamma.Q} = Z_2 + \frac{P_2}{\gamma} + \frac{V_2^2}{2g} + h_p \tag{1}$$

The variable diameters used in the Bernoulli equation for the conservation of energy in fluids are illustrated in Fig. 5. In a Bernoulli equation, one of the analysis points is in the upper part of the fluid inside the tank and the other at the outlet of the tank and the electric pump.

After the linearization and subsequent application of the Laplace transform, the transfer function of the system—G(S) in Fig. 6—can be obtained, which can be

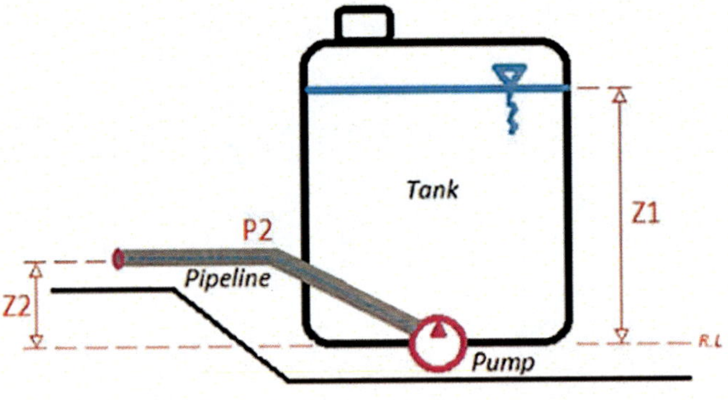

Fig. 5 Application of Bernoulli's principle in the system

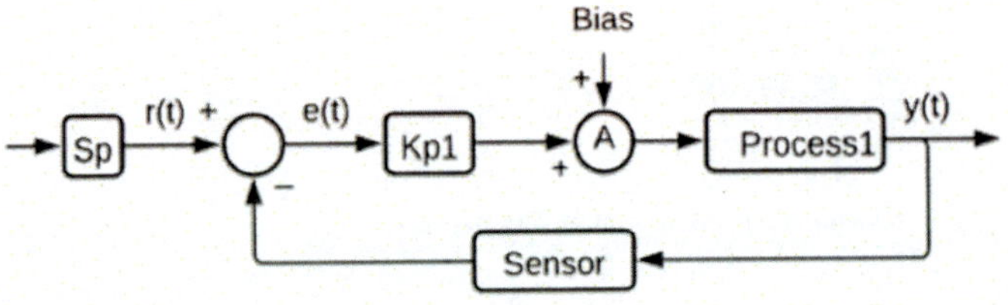

Fig. 6 Block diagram of the fluid pressure control system

manipulated with the electric pump controller, leaving only this to apply the constant of proportionality, Kp1, to stabilize the system response.

3 Results

3.1 System Integration

The design features an electronic flow control system on a mobile mechanical structure that will traverse fields via remote operation, as shown in Fig. 7. It opens or closes the solenoid valves in proportion to plants' location and size. This design has a control circuit that can be controlled remotely, via either a cell phone or any other electronic device equipped with a Bluetooth or Wi-Fi connection, in order to prevent the operator from being exposed to agrochemicals during aspersion.

The operation of the sprinkler control system takes place in the following steps:

First, the reference points must be entered manually. They must be in ascending order, as can be seen in Fig. 8, which serves as a working range for how open the solenoid valves should be. These values can be configured from 20 to 60 cm, where the maximum height of the valves is also the maximum height of the structure.

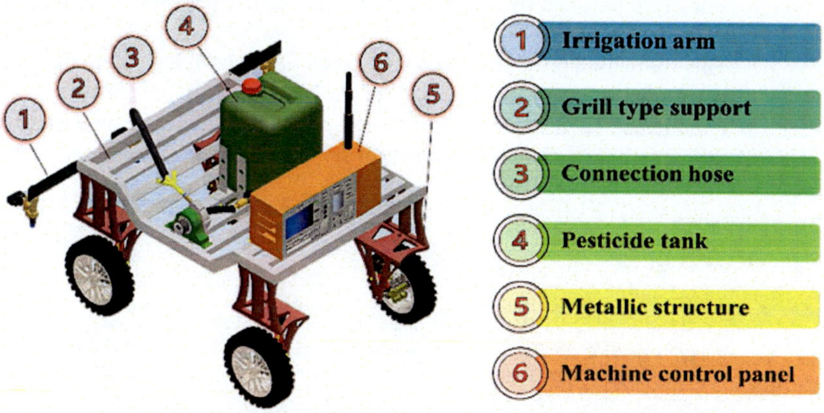

Fig. 7 Integration of the joint mechatronic system

Fig. 8 How much the solenoid valves open, which corresponds to the height of plants

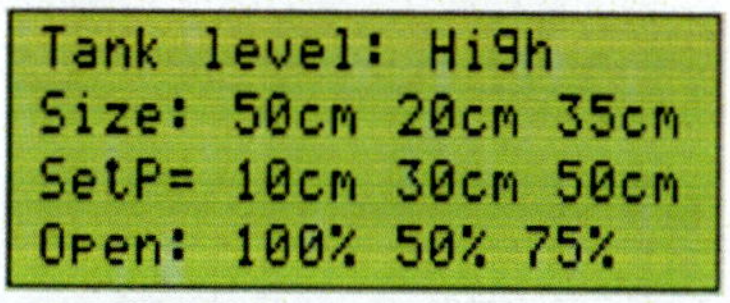

Second, upon detecting a plant, the system calculates the plant's approximate height so that the system can locate the plant within its range. Previously established set points determine when it will proceed with opening the valves in proportion to the calculated height.

Finally, as the output of the system, the proportional opening of the solenoid valves was obtained from previously processed data. The maximum valve opening angle is 90°, and a rule of three is followed to set values for the opening angles of the valves. These angles increase or decrease depending on each plant's requirements, which are themselves determined by each plant's location and position at specific times. The amount of spray delivered to each plant also directly corresponds to the speed at which the vehicle moves.

4 Conclusion

A combination of the correct dosage of agrochemicals and the correct calibration of an automatic sprinkler system adequately controlled pests. Optimizing the dosage in this way reduced the amount of pesticide contamination released into the environment, improved cost-effectiveness, and reduced farmers' risks of being poisoned because they no longer had direct contact with these chemicals. Therefore, this research proposed the design of a proportional spraying control system using a remotely controlled mobile robotic system for the delivery of agrochemicals to crops.

This study used proportional control because the sprinkler system must be able to respond quickly to any displacement of the mobile structure. This study developed and implemented a remotely controlled robotic system and an electronic flow control system, which will be further tested and analyzed in future research.

References

1. Food and Agriculture Organization of the United Nations FAO. (2014). *Family farming in Latin America and the Caribbean: Policy recommendations*. E-ISBN 978-92-5-308364-0.
2. World Health Organization. *Pesticide residues in food*. [Online]. Available at: https://www.who.int/es/news-room/fact-sheets/detail/pesticide-residues-in-food
3. Jiménez, A. O., Rodríguez, F. S., & Castuera, M. H. (2011). Estimation of the degree of knowledge about the chemical risk in Badajoz workers. *Journal of Toxicology, 28*(2), 158–165.
4. Calaf, G. M. (2021). Role of organophosphate pesticides and acetylcholine in breast carcinogenesis. In *Cancer biology seminars* (Vol. 76, pp. 206–217). Academic Press.
5. Ba Hashwan, S. S., Khir, M. H. B. M., Al-Douri, Y., & Ahmed, A. Y. (2020). Recent progress in the development of biosensors for chemicals and pesticides detection. *IEEE Access, 8*, 82514–82527. https://doi.org/10.1109/ACCESS.2020.2991380
6. Ministry of Health MINSA. *Health situation room. Epidemiological surveillance of the risk of exposure and poisoning by pesticides*. Available at: https://www.dge.gob.pe/portal/docs/tools/Vplaguicidas040718.pdf

7. Dong, Y., et al. (2020). Automatic system for crop pest and disease dynamic monitoring and early forecasting. *IEEE Journal of Selected Topics in Applied Earth Observations and Remote Sensing, 13*, 4410–4418. https://doi.org/10.1109/JSTARS.2020.3013340
8. World Bank. *Agriculture and food.* [Online]. Last update: September 23, 2019. Available at: https://www.bancomundial.org/es/topic/agriculture/overview
9. Huachos, H., & Maribel, K. (2017). *The use of chemical pesticides in potato crops (Solanum Tuberosum L), Its relationship with the environment and health.*
10. Mogili, U. R., & Deepak, B. B. V. L. (2018). Review on application of drone systems in precision agriculture. *Procedia Computer Science, 133*, 502–509.
11. Grimstad, L., & From, P. J. (2017). Thorvald II-a modular and re-configurable agricultural robot. *IFAC-PapersOnLine, 50*(1), 4588–4593.
12. Razvarz, S., Vargas-Jarillo, C., Jafari, R., & Gegov, A. (2019). Flow control of fluid in pipelines using PID controller. *IEEE Access, 7*, 25673–25680. https://doi.org/10.1109/ACCESS.2019.2897992
13. Chen, T., & Meng, F. (2018). Development and performance test of a height-adaptive pesticide spraying system. *IEEE Access, 6*, 12342–12350. https://doi.org/10.1109/ACCESS.2018.2813667
14. Minov, S. V., Cointault, F., Vangeyte, J., Pieters, J. G., & Nuyttens, D. (2015). Pesticide spray characterization using high speed imaging techniques. *Agriculture and Agricultural Science Procedia, 7*, 280–286. https://doi.org/10.1016/j.aaspro.2015.12.048
15. Gausemeier, J., & Moehringer, S. (2002). VDI 2206-a new guideline for the design of mechatronic systems. *IFAC Proceedings Volumes, 35*(2), 785–790.
16. Rivera, G., et al. (2019). Mechatronic device for the analysis and mitigation of involuntary movements in people with Parkinson's disease. *UTE Approach, 10*(1), 153–172.
17. Gómez, M., Pedro, F., et al. (2019). Automation of greenhouse pesticide application: Design and construction. *Electronic Vision, 2*(1), 129–133.
18. Medina, C. E. G., Hernández, D. A. B., & Alejandro, D. (2018). *Knapsack sprayer with electronic flow control.*
19. Onorato, A., & Tesouro, M. O. (2004). Anti-drift performance of an air-induced hollow cone agricultural spray nozzle. ESTUARY. *Revista de Investigaciones Agropecuarias, 33*(3), 3–13.
20. González, V., Cristian, O., et al. (2017). Bluetooth technologies applied to the control of unmanned land vehicles for spraying agrochemicals on ornamental plants. *Scientific, 21*(2), 127–134.

Index

© The Editor(s) (if applicable) and The Author(s), under exclusive license to Springer Nature Switzerland AG 2024
S. Yeom (ed.), *4th International Conference on Electronics and Signal Processing*, Signals and Communication Technology,
https://doi.org/10.1007/978-3-031-36670-3